D0908723

COMPOSTING MUNICIPAL SLUDGE

COMPOSTING MUNICIPAL SLUDGE

A Technology Evaluation

by

Arthur H. Benedict

Brown and Caldwell
Pleasant Hill, California

Eliot Epstein and Joel Alpert

E&A Environmental Consultants, Inc.
Stoughton, Massachusetts

NOYES DATA CORPORATION
Park Ridge, New Jersey, U.S.A.

Library of Congress Catalog Card Number 87-34746
ISBN: 0-8155-1162-0
ISSN: 0090-516X
Printed in the United States

Published in the United States of America by
Noyes Data Corporation
Mill Road, Park Ridge, New Jersey 07656

10 9 8 7 6 5 4 3 2 1

Library of Congress Cataloging in Publication Data

Benedict, Arthur H.
Composting municipal sludge : a technology evaluation / by Arthur H. Benedict, Eliot Epstein, and Joel Alpert.
p. cm. -- (Pollution technology review, ISSN 0090-516X ; no. 152)
Bibliography: p.
Includes index.
ISBN 0-8155-1162-0
1. Sewage sludge as fertilizer--United States. 2. Compost--Evaluation. 3. Sewage--Purification--Activated sludge process--Evaluation. I. Epstein, Eliot, 1929- . II. Alpert, Joel E. III. Title. IV. Series
TD772.B46 1988
628.3'6--dc19 87-34746
CIP

Foreword

The objectives of this study were to assess the design, construction, operation, and cost features of static pile and windrow technologies for municipal sludge composting applications. Three static pile facilities, one conventional windrow facility, and one aerated windrow facility were investigated.

Widespread interest in composting as a means of municipal sludge treatment in the United States began in the early 1970's. At that time, the Los Angeles County Sanitation Districts initiated windrow composting of sewage sludge at the Joint Water Pollution Control Plant in Carson, California, and the U.S. Department of Agriculture investigated large-scale studies of static pile composting at the Agricultural Research Station in Beltsville, Maryland. Since that time, interest and activity in municipal sludge composting has increased dramatically.

In 1984, the U.S. Environmental Protection Agency (EPA) initiated a technology evaluation of municipal sludge composting practice based on investigations at five operating facilities. The results of the technology evaluation are presented in the book and summarized below.

Static pile and windrow technologies typically require 0.08 to 0.14 acre per wet ton per day. Paving and covering key operating areas can enhance reliability, but site-specific factors need to be assessed. Flexibility to respond to variable sludge loadings and moisture control is a major requirement for effective performance. Odor control is the main off-site issue that must be addressed during design and operation. Both technologies can produce a marketable product under variable operating conditions.

The information in the book is from *Composting Municipal Sludge: A Technology Evaluation* prepared by Arthur H. Benedict of Brown and Caldwell and by Eliot Epstein and Joel Alpert of E&A Environmental Consultants, Inc. for the U.S. Environmental Protection Agency, March 1987.

The table of contents is organized in such a way as to serve as a subject index and provides easy access to the information contained in the book.

Advanced composition and production methods developed by Noyes Data Corporation are employed to bring this durably bound book to you in a minimum of time. Special techniques are used to close the gap between "manuscript" and "completed book." In order to keep the price of the book to a reasonable level, it has been partially reproduced by photo-offset directly from the original report and the cost saving passed on to the reader. Due to this method of publishing, certain portions of the book may be less legible than desired.

Acknowledgments

Many individuals contributed to the preparation and review of this report and the field investigations that served as a primary basis of this project. Contract administration was provided by the Water Engineering Research Laboratory (WERL) of the Office of Research and Development of the U.S. Environmental Protection Agency (EPA) in Cincinnati, Ohio.

Contractor-Authors:

Brown and Caldwell, Consulting Engineers, Pleasant Hill, CA
Arthur H. Benedict, Project Manager

E&A Environmental Consultants, Inc., Stoughton, MA
Eliot Epstein
Joel Alpert

Contract Supervision:

U.S. Environmental Protection Agency, WERL, Cincinnati, OH
James F. Kreissl, Project Officer
Donald S. Brown, Project Manager

Subcontractors:

E&A Environmental Consultants, Inc., Stoughton, MA
Robert E. Lee & Associates, Inc., Green Bay, WI

Field Investigations:

Performance of the field investigations required the participation of many individuals and municipalities. Field investigations were conducted at composting facilities in Newport News, VA; Silver Spring, MD; Columbus, OH; Los Angeles, CA; and Denver, CO. These investigations could not have been completed without the commitment of personnel from these facilities. The assistance of the following individuals is therefore gratefully acknowledged:

Hampton Roads Sanitation District Peninsula Composting Facility, Newport News, VA
Mardane McLemore
Todd O. Williams
Danny L. Finley

Washington Suburban Sanitary Commission Montgomery County Composting Facility, Silver Spring, MD
Charles M. Murray
George O. Crosby
Joel A. Thompson

City of Columbus, OH, Southwesterly Composting Facility, Franklin County, Ohio
Donald R. Rodgers
Duane E. Goodridge

Los Angeles County Sanitation Districts Joint Water Pollution Control Plant Composting Facility, Carson, CA
Charles Carry
Ross C. Caballero

Metropolitan Denver Sewage Disposal District Number One Demonstration Composting Facility, Denver, CO
William Martin
Marvin Webb

NOTICE

Contents and Subject Index

1. Introduction

Widespread interest in composting as a means of municipal sludge treatment in the United States began in the early 1970s. At that time, the Los Angeles County Sanitation Districts initiated windrow composting of sewage sludge at the Joint Water Pollution Control Plant in Carson, California, and the U.S. Department of Agriculture investigated large-scale studies of static pile composting at the Agricultural Research Station in Beltsville, Maryland. Since that time, interest and activity in municipal sludge composting has increased dramatically.

In 1984, the U.S. Environmental Protection Agency (EPA) initiated a technology evaluation of municipal sludge composting practice based on investigations at five operating facilities. The results of the technology evaluation are presented in this report.

OBJECTIVES

Objectives of the municipal sludge composting technology evaluation were as follows:

1. To investigate aerated static pile and windrow composting technologies based on experience at operating facilities.

2. To compare and contrast features of the aerated static pile and windrow technologies based on this experience.

3. To assess operating, performance, and cost features.

4. To identify key problems associated with municipal sludge composting using these technologies.

5. To define methods which have been used or are being considered to resolve these problems.

The technology evaluation focused on three composting processes: the extended aerated static pile process, the conventional windrow process, and the aerated windrow process. In-vessel, mechanical processes were not evaluated.

FACILITIES INVENTORY

The municipal sludge composting technology evaluation was initiated with an inventory of operating facilities in the United States. The inventory was prepared by identifying operating facilities in each state and making telephone inquiries regarding current operations. The complete inventory is presented as Appendix A and a summary is presented in Table 1.[1]

A total of 42 locations composting solely municipal sludge were identified as of spring 1984. Twenty-eight utilize the aerated static pile process, twelve use conventional windrow techniques, and two employ aerated windrowing. Most of the facilities are under 10 dry tons per day (dtpd) and have been operating about 5 years or less.

A variety of sludges are processed by the facilities, including those which are undigested (raw), anaerobically digested, and aerobically digested. Ten of the aerated static pile facilities inventoried report representative dewatered sludge total solids concentrations consistently below 15 percent, and five of the conventional windrow systems are in this category. Thirteen static pile facilities report dewatered solids typically in the 15 to 20 percent range, although loads having dewatered solids under 15 percent are occasionally received. Five of the conventional windrow and one of the aerated windrow facilities report dewatered sludge total solids concentrations that are in this category. Eight of the municipal sludge composting facilities report dewatered sludge total solids concentrations consistently greater than 20 percent. One of these, an aerated windrow facility, receives a heavily limed sludge having a total solids concentration of about 40 percent.

COMPOSTING TECHNOLOGIES

Schematics for the aerated static pile, conventional windrow, and aerated windrow composting processes are presented on Figure 1. The aerated static pile process involves mixing dewatered sludge with a bulking agent, such as wood chips, followed by active composting in specially constructed piles such as shown on Figure 2. Typically, both recycled bulking agent and new (external) bulking agent are used for mixing. Induced aeration, either positive (blowing) or negative (suction), is provided during the active composting phase. Temperature and oxygen are monitored during active composting as a means of process control. The active composting period lasts at least 21 days, following which alternate pathways to produce finished compost may be employed as described below.

If at the end of the 21-day active composting period, composted material is sufficiently dry, screening may be performed directly to separate bulking agent for recycle. The recycled bulking agent is generally stored prior to reuse in the mixing operation. Screened compost is restacked and cured for at least 30 days and then stockpiled as finished compost prior to distribution.

If at the end of the 21-day active composting period, compost material is not sufficiently dry for screening, a separate drying step is required

[1]Only facilities composting solely municipal sludge were inventoried. Those composting sludge in conjunction with refuse or other waste materials were not included.

TABLE 1. INVENTORY OF OPERATING MUNICIPAL SLUDGE COMPOSTING FACILITIES, SPRING 1984

Facility characteristics	Facility composting process		
	Aerated static pile	Conventional windrow	Aerated windrow
Total number	28	12	2
Nominal size, dry tons per day			
Less than 10	19	9	1
10 to 25	3	2	1
26 to 50	3	0	0
51 to 100	2	0	0
Greater than 100	1	1	0
Years of operation			
Less than 1	3	0	0
1 to 5	21	7	1
6 to 10	4	1	1
Greater than 10	0	4	0
Sludge total solids, percent			
Less than 15	10	5	0
15 to 20	13	5	1
Greater than 20	5	2	1

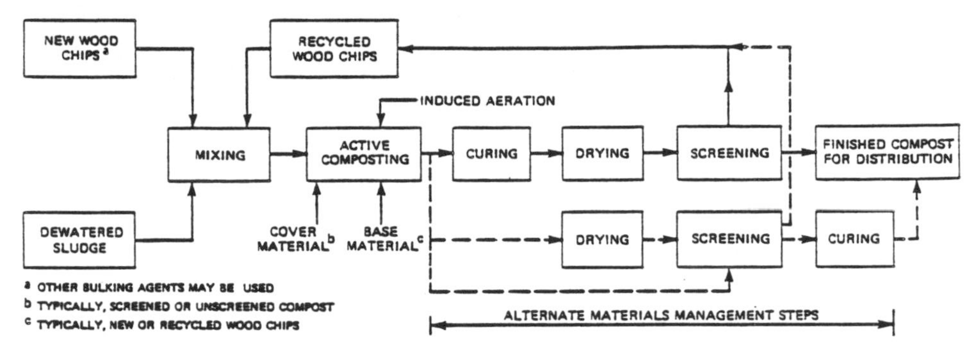

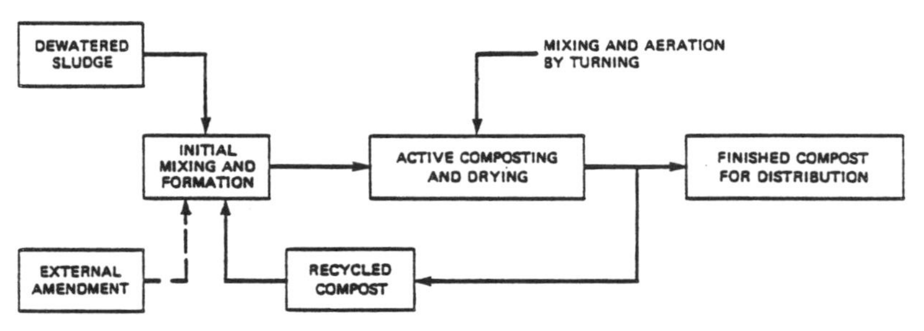

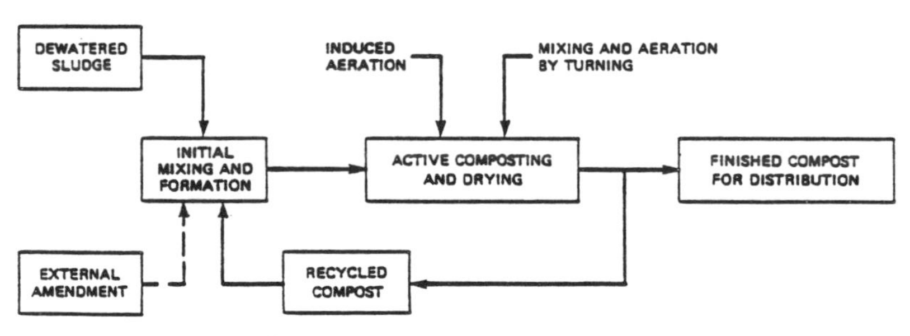

FIGURE 1. SLUDGE COMPOSTING PROCESS SCHEMATICS

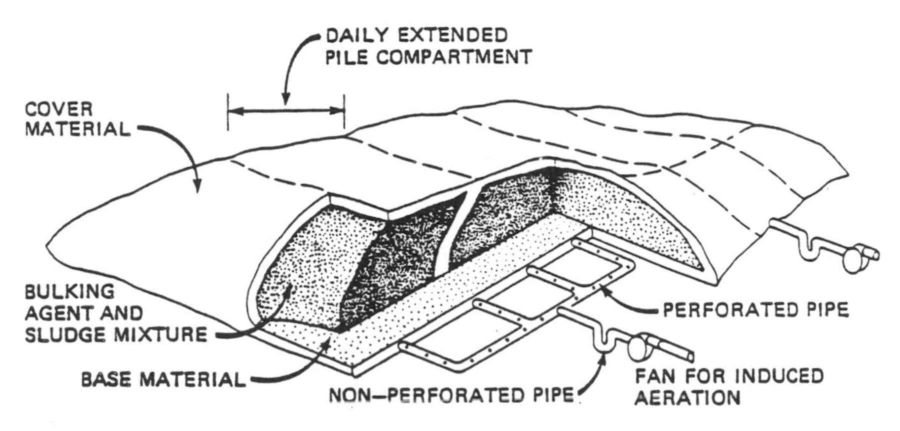

A. EXTENDED AERATED STATIC PILE COMPOSTING

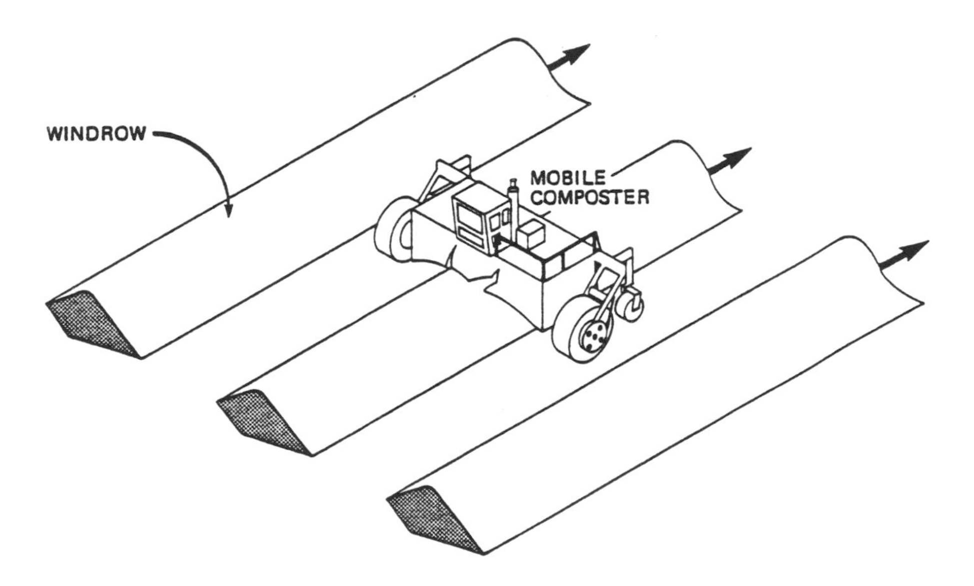

B. CONVENTIONAL WINDROW COMPOSTING

FIGURE 2. SLUDGE COMPOSTING METHODS

prior to screening, curing, and stockpiling. Alternatively, unscreened compost may be restacked for the 30-day curing period, after which drying, screening, and stockpiling are performed.

The conventional windrow process (Figure 1) involves initial mixing of dewatered sludge with a bulking agent such as finished compost, often supplemented with an external amendment, followed by formation of long windrows such as shown on Figure 2. Formation of the windrows is generally performed in two steps. Typically, front-end loaders are used to initially stack material in a rough windrow configuration; then a specially designed mobile composter is used to fine mix the material by turning the windrow in place.

An active windrow composting period of 30 days (or more) is provided following initial mixing and formation. During this period, the windrows are periodically turned with a mobile composter (in some cases front-end loaders maybe used) to aerate and remix the material. A turning frequency of two to three times per week is typical. Temperature is monitored for process control. Following the active windrow composting period, the composted material is allowed to cure for at least 30 days; then, a portion of the finished compost is recycled and a portion is stockpiled for distribution.

The aerated windrow process is similar to the conventional windrow process with one exception: A system for induced aeration is provided in addition to aeration by turning with a mobile composter. Either positive- or negative-induced aeration may be used, which is intended to enhance active composting and drying. Typically, induced aeration is provided using a fan and a fixed arrangement of pipes or channels for delivering air uniformly to the base of the windrow. Details of the air delivery system used at one of two operating aerated windrow facilities presented in Table 1 are presented subsequently in this report.

FACILITIES AND METHODS

The municipal sludge composting technology evaluation focused on operating, performance, and cost features of the aerated static pile, conventional windrow, and aerated windrow composting processes based on investigations at five operating facilities: the Hampton Roads Sanitation District Peninsula Composting Facility (Hampton Roads facility) located in Newport News, Virginia; the Washington Suburban Sanitary Commission Montgomery County Composting Facility (Site II facility) located in Silver Spring, Maryland; the City of Columbus, Ohio, Southwesterly Composting Facility (Columbus facility) located in Franklin County, Ohio; the Joint Water Pollution Control Plant Composting Facility of the Los Angeles County Sanitation Districts (Los Angeles facility) located in Carson, California; and the Metropolitan Denver Sewage Disposal District Number One Demonstration Composting Facility (Denver facility) located in Denver, Colorado. Key features of the five composting facilities investigated during the technology evaluation are presented in Table 2.

The Hampton Roads facility is an aerated static pile operation processing a mixture of anaerobically digested primary and waste activated sludge. Two aerated static pile facilities process raw sludge: The Site II facility

TABLE 2. KEY FACILITY FEATURES

Facility name and location	Composting process	Sludge characteristics
Hampton Roads Sanitation District Peninsula Composting Facility, Newport News, Virginia	Static pile	Anaerobically digested primary and secondary
Washington Suburban Sanitary Commission Site II Composting Facility, Silver Spring, Maryland	Static pile	Limed, raw primary and secondary
City of Columbus, Ohio, Southwesterly Composting Facility, Franklin County, Ohio	Static pile	Unlimed, raw primary and secondary
Los Angeles County Sanitation Districts Joint Water Pollution Control Plant Composting Facility, Carson, California	Conventional windrow	Anaerobically digested primary and secondary
Metropolitan Denver Sewage Disposal District Number One Demonstration Composting Facility	Aerated windrow[a]	Anaerobically digested primary and secondary

[a]The conventional windrow process was also utilized as part of a composting demonstration study at this facility.

treats limed, raw sludge, and the Columbus facility treats unlimed, raw sludge. The only conventional windrow operation investigated was the Los Angeles facility which processes anaerobically digested primary and secondary sludge. Similarly, the Denver facility is the only aerated windrow facility investigated. It was designed for demonstration study purposes to treat anaerobically digested primary and secondary sludge and also utilized the conventional windrow process as part of the study.

The scope of the technology evaluation varied at each composting facility. At the three static pile facilities, on-site investigations which consisted of a review of plant records and observations of facility operations were conducted. Independent process testing was also performed at the Hampton Roads and Columbus facilities but not at the Site II facility. On-site studies at the Hampton Roads and Columbus facilities lasted 3 weeks each, whereas the on-site study at the Site II facility was conducted over a 4-day period. The investigation of conventional windrow composting at the Los Angeles facility was based on a 1-day site visit and a review of key published information provided by facility personnel, whereas that at the Denver aerated windrow facility was based on a 4-day site visit. No independent process testing was performed at either of the windrow facilities.

REPORT CONTENTS

The remainder of this report is organized into seven chapters. Chapter 2 presents the conclusions and recommendations of the technology evaluation. Descriptions of the three static pile facilities and the two windrow facilities investigated during the technology evaluation are presented in Chapters 3 and 4, respectively. An operations assessment for static pile composting is presented in Chapter 5 and that for windrow composting is presented in Chapter 6. A cost evaluation is presented in Chapter 7, and a comparison of the static pile and windrow composting technologies is provided in Chapter 8.

2. Conclusions and Recommendations

Conclusions and recommendations of the municipal sludge composting technology evaluation are presented in this chapter. The conclusions drawn and recommendations offered are based on investigations at three aerated static pile facilities, one conventional windrow facility, and one aerated windrow facility. The aerated windrow facility was developed and operated for demonstration study purposes, whereas the others are fully operational. Facilities employing in-vessel, mechanical composting technologies are not included in the evaluation. The facilities investigated vary in size, type of sludge composted, geographic location, climatic conditions, and such site-specific factors as surrounding land use characterisitcs, market potential for finished compost, and type of equipment and operating procedures applied.

The technology evaluation was performed to assess a broad range of municipal sludge composting activities related to design, operation, performance and economics at each of the facilities studied. The report is therefore intended to be used as an information source on municipal sludge composting rather than as a design manual.

CONCLUSIONS

General conclusions of the municipal sludge composting technology evaluation and conclusions specific to static pile or windrow technologies are presented below.

General

Conclusions based on a comparison of static pile and windrow composting technologies are presented in this section.

Moisture Control--

Mosture control and a homogeneous initial mix were found to be the most important factors for effective composting. During day-to-day operations moisture must be controlled for effective stabilization, pathogen inactivation, odor control, and finished compost quality control. The moisture can come from several sources including: the incoming sludge, bulking agent, or amendment; inclement weather; and the composting process itself as a by-product

An initial mix moisture content of 60% or less (total solids >=40%) was found to be a key criterion regardless of the composting technology employed.

To be a marketable product the final compost must generally have 50 to 60 percent total solids.

Sludge Loading Variability--

Sludge loadings for design must be carefully established and must consider the high variability of both the quantity and quality of sludge to be processed. For example, peak-to-average ratios for loadings at the static pile facilities studied ranged from 1.4 to 1.9, and the total solids solids content of incoming sludge at these facilities ranged from 14% to 22%. Higher than anticipated values for moisture content, average daily loading, or peak day loading can cause several operational problems. Bulking agent or amendment use, mixing operations, personnel utilization, equipment utilization and effectiveness, area requirements, and composting performance can all be adversely impacted by loading variations. Therefore, operational flexibility to respond to the day-to-day variability in sludge total solids (moisture) content and the wet quantity of sludge received must be considered during design.

Although loadings based on calendar days are useful for general comparisons, operating-day loadings (for example, based on 5-day-, 6-day-, or 7-day-per-week operation) more accurately reflect day-to-day materials management and process requirements, and should be used for design purposes.

Wastewater treatment plant operations can also impact the effectiveness of composting operations. Variability in dewatered sludge total solids caused by changes in feed sludge characteristics to the dewatering operation is one example. Thus, integrating treatment plant and composting operations is important for effective composting performance. Timing sludge deliveries, coordinating plant operations with composting constraints, and providing sludge storage are examples of ways in which this can be accomplished.

Operational Area--

Unit operational area for static pile and windrow composting, expressed as acres per wet ton per day (ac/wtpd), is site-specific. Operational area is defined as site area required for process operations and support facilities such as administrative and maintenance buildings, but excluding extraneous areas such as land held on site for expansion purposes. Based on the facilities investigated, a reasonable range is 0.08 to 0.14 ac/wtpd, depending on such site-specific considerations as daily sludge loadings, storage requirements, drying technique employed, composting method, and the need for runoff control ponds.

Paving and Covering--

All of the facilities investigated have paved areas for most of their operations. Operating reliability is enhanced by the use of paved surfaces, based on wet-weather problems which have been experienced at the facilities studied. Covering or enclosing key operating areas can also enhance reliability; for example, all of the static pile facilities have protected mixed areas. However, site-specific and technology-specific factors are important in defining the need for protection at any given facility.

Equipment and Manpower Utilization--

Front-end loaders are used for a number of static pile and windrow operations, including rough mixing, pile construction, windrow formation, teardown activities, and materials storage or transfer operations. Mobile composters are used for fine mixing during static pile composting as well as for windrow turning.

Generally, manpower utilization is site-specific. However, based on the number of operating personnel employed at the facilities studied, the static pile technology is more labor-intensive than the conventional windrow technology.

Odor Control--

Odor generation and release is the main environmental problem at static pile and windrow composting installations, and odor management is a complex problem. A certain amount of odor is unavoidable, and a good public relations campaign should be used to minimize public opposition. However, some design and operational features can be employed to minimize the potential for generation or release of objectionable odors, including the following:

1. Trucks used to haul dewatered sludge should be covered and cleaned frequently. This is particularly important with raw sludge.

2. Dewatered sludge deliveries should be managed such that mixing and other operations can be performed without sludge accumulating for long periods of time. This is particularly important with raw sludge and in hot weather.

3. An initial mix moisture content of 60 percent or less is a key criterion for odor control. Enclosing operations and scrubbing exhaust gas can be an effective step for odor control in some situations.

4. A uniformly mixed and sufficiently porous material is important for odor control during composting. The presence of clumps of unmixed sludge can lead to anaerobic and, thus, odorous conditions, as well as incomplete stabilization and pathogen inactivation. Construction of daily static pile compartments, with cover material, is also important.

5. Simply limiting the daily quantity of sludge composted at a facility has been used as an odor control technique.

6. Positive aeration during active static pile composting utilizes the pile cover material as an odor scrubber. Negative aeration requires the use of a separate exhaust scrubber system such as a finished compost filter pile. Regardless of the aeration mode employed, aeration rate, oxygen content, and temperature must be controlled to promote active aerobic composting.

7. During the life of a conventional windrow, continuous emissions from the windrow surface are a greater source of odor than the periodic emissions during windrow turning. Although emissions are most intense immediately after turning, they then decrease rapidly. Surface emissions decrease as the composting process progresses.

8. Teardown of static piles or windrows can be managed to minimize release of odors. For example, teardown can be restricted during wet weather or during early morning hours when air inversions may occur. At static pile installations, high-rate aeration for 24 to 48 hours prior to teardown, which lowers pile temperature to ambient, can be used to control odor release.

9. Leachate, condensate, and runoff should be collected and disposed to minimize odor generation potential. Proper site drainage is required to prevent ponding which can generate odor.

10. Effective housekeeping procedures such as washing equipment and flushing or sweeping working areas also reduce odor generation potential.

Performance--

Both static pile and windrow composting meet pathogen inactivation and volatile solids reduction requirements and produce a marketable product while operating under a wide range of conditions. Sustained finished compost production between 0.4 and 1 cubic yard per wet ton (cu yd/wt) of sludge composted is representative of the facilities investigated. This production rate is equivalent to about 0.7 to 2 dry tons of compost per dry ton of sludge. (The apparent increase in mass is due to the incorporation of wood chip fines or amendment material into the finished compost.)

Facility Costs--

Representative costs in 1985 dollars for static pile composting are $36 to $51 per wet ton of sludge based on the facilities studied. These costs are higher than the costs estimated at design. Amortized capital costs based on a 20-year amortization period and an interest rate of 10 percent, and annual operation and maintenance costs, are included in these figures. Dewatering and sludge transport costs are excluded. Comparative costs for windrow technologies were not established during the technology evaluation. Operating cost reductions equivalent to $1 to $3 per wet ton are currently achieved from compost sales revenues at the facilities studied and all have marketing programs to increase revenues in the future.

Static Pile Composting

Conclusions specific to the static pile facilities studied are presented in this section.

Wood Chip Usage--

Dewatered sludge total solids content being received at each static pile facility studied is below that estimated at design by 3 to 5 percent. Because of this, wood chip use is as much as 80 percent over the design estimates. Increased wood chip use has increased composting and storage area requirements, constricted equipment mobility, and increased operating cost. Typical mix ratios are between 3.5 and 4.5 cu yd/wt, depending on season, proportion of new and recycled chips, and chip moisture content.

Initial Mixing--

In the absence of a standard test method, the use of a single person to oversee mixing operations ensures consistent and proper initial mix moisture content and mix homogeneity. Having more than one person responsible for mixing adversely impacts the effectiveness of the mixing operation and therefore should be avoided. A mixing time of 40-45 is typically required.

Pile Construction--

Extensive testing following initial construction was required at two of the static pile facilities to arrive at reliable pile construction and aeration piping configurations for routine operation. Both uniform and variable hole spacing in the aeration piping is used. Extended static pile heights are typically 12 to 13 ft, but lengths and widths of daily pile compartments vary based on daily loadings. Minimum depths of base and cover materials are 12 and 18 inches, respectively. Two of the facilities use denser material (unscreened compost or dry, recycled compost) at the pile toes to prevent short-circuiting of air flow.

Composting Period--

A minimum active composting period of 21 days is required for effective composting at static pile facilities, but longer periods are sometimes used because of materials management considerations. Regulations require that internal temperatures at all monitoring points be maintained at >=55 degrees C for at least 3 days for pathogen inactivation and degradation of putrescrible organics.

Process Control--

Variation of aeration rates to achieve desired temperature and oxygen levels during active composting is the key process control method. Aeration rates per dry ton, number of blowers per pile compartment, individual blower capacity, total blower capacity on site, and blower capacity per wtpd vary from site to site. However, the trend is now toward providing one blower per daily pile compartment. Generally speaking, aeration rate is controlled by cycling blower operation until the time-temperature regulations are met. After this, high-rate, continuous aeration is used to lower the temperature and promote drying. Manual and automatic blower control are used, and both positive and negative aeration are used.

Drying--

Material composted by the aerated static pile method is usually not dry enough to screen directly, and a separate drying step is employed. A minimum unscreened compost solids content of 50 to 55 percent is generally required for effective screening. All of the static pile facilities investigated use some type of separate drying step and also use high-rate, continuous aeration during the final stage of active composting to promote drying.

Screening--

Screening effectiveness, including expected variability, needs to be carefully assessed in aerated static pile applications. Under optimum conditions, wood chip recoveries equivalent to 80 to 90 percent (based on unscreened compost volume) can be obtained. However, this requires that unscreened compost total solids content be between 50 and 60 percent, depending, in part, on screen size and design. Exceeding the upper limit can result in losses from excessive dust generation, while not meeting the lower limit inhibits the separation of fines from wood chips and leads to wood chip contamination, reduced screening rates, and often mechanical problems with the screening equipment. Wood chip recoveries of 65 to 85 percent are more typical of sustained routine operation at the static pile facilities investigated.

Curing--

Curing operations at static pile facilities are dependent, in large part, on materials management requirements. A minimum curing period of 30 days is provided, but longer periods are often used. Unscreened or screened compost curing can be used, but an advantage of screened compost curing is that it requires less area than unscreened compost curing. Aerated curing can also be effective.

Performance--

Finished compost having a total solids content of 51 to 69 percent and bulk densities of 900 to 1,500 pounds per cubic yard, and free of Salmonella is routinely produced for sale at static pile facilities. Other characteristics, particularly metal content, are site-specific.

Operating Cost--

On-site labor, wood chip purchase, and aeration pipe replacement account for about 50 to 80 percent of the annual operation and maintenance (O&M) cost at static pile facilities. The annual O&M cost in 1985 dollars ranged from $25 to $36 per wet ton of sludge at the facilities studied.

Windrow Composting

Conclusions specific to the windrow facilities studied are presented in this section.

Windrow Formation--

Initial rough mixing of sludge and amendment prior to fine mixing and windrow shaping using mobile composters is required for effective windrow composting. Equipment used for initial rough mixing varies.

Composting Period--

A minimum active composting period of 30 days is required for windrow composting, but longer periods are sometimes used. Regulations require that internal temperatures at all monitoring points be maintained at >=55 degrees C for at least 15 days. Use of large windrows with low surface-to-volume ratios (7 ft high and 23 ft wide at the base) during the latter stage of composting can conserve heat and aid temperature development for pathogen inactivation and degradation of putrescible organics. A turning frequency of two to three times per week is adequate for mixing and aeration. An additional curing step is not required.

Drying--

Based on an independent analysis of demonstration study results from the aerated windrow facility, induced aeration can improve drying by about 3 percent over conventional windrowing. The type of amendments used can also aid drying in some instances.

Dust Generation--

Dust generated from mixing, windrow turning, and sometimes windrow surfaces can be a problem.

Performance--

Windrow composting can routinely produce finished compost having a total solids content of up to 60 percent and free of Salmonella.

Operating Cost--

Labor cost is the main operation and maintenance cost at windrow compost facilities.

RECOMMENDATIONS

Based on the findings of the municipal sludge composting technology evaluation, it is recommended that:

1. Further studies be performed on the composting technologies investigated to refine design and operating criteria needed to optimize moisture and odor control.

2. An update of existing process design manuals for municipal sludge composting be prepared, incorporating recent experience at operating facilities such as those investigated during the technology evaluation.

3. Guidelines for start-up of composting facilities be developed. This is particularly important for static pile facilities to arrive at reliable pile construction techniques, aeration configurations, and screening reliability.

4. A technology evaluation of in-vessel, mechanical composting technologies be performed to complement the evaluation of static pile and windrow technologies.

3. Static Pile Facility Descriptions

Three aerated static pile composting facilities were investigated during the municipal sludge composting technology evaluation: the Hampton Roads Sanitation District Peninsula Composting Facility (Hampton Roads facility) located in Newport News, Virginia; the Washington Suburban Sanitary Commission Montgomery County Composting Facility (Site II facility) located in Silver Spring, Maryland; and the City of Columbus, Ohio, Southwesterly Composting Facility (Columbus facility) located in Franklin County, Ohio. A general description of each facility and a comparison of static pile site features is presented in this chapter. Additional descriptive information is included in the static pile operations assessment presented in Chapter 5.

HAMPTON ROADS FACILITY

The planning background, design considerations, and general features for the Hampton Roads facility are presented in this subsection.

Facility Planning and Design

The Hampton Roads facility utilizes the extended aerated static pile process to compost anaerobically digested primary and secondary sludge. Facility planning was initiated in 1979 to provide for disposal of sludge at the James River and York River wastewater treatment plants over a 13-year period (1979 to 1992). Construction was completed and the facility began operation in October 1981.

Initial planning for the Hampton Roads facility analyzed incineration, landfilling, and composting alternatives at the James River plant and landfilling and composting at the York River plant. Based on a cost-effectiveness analysis of monetary and nonmonetary factors, a centralized facility to serve both plants was selected.

The Hampton Roads facility currently processes anaerobically digested sludge from a third location, the Nansemond wastewater treatment plant. Composting of the Nansemond sludge is on an interim basis only and will be phased out as loadings from the James River and York River facilities increase to design levels. All three treatment plants employ primary sedimentation followed by air activated sludge treatment. Belt filter presses with polymer addition are used for sludge dewatering.

The Hampton Roads facility was designed to meet State of Virginia regulations for the application of municipal sludge to croplands. These regulations designate maximum allowable metal levels and require that sludge be treated by a process to further reduce pathogens (PFRP). For the aerated static pile composting process, compost temperatures must reach 55 degrees centigrade (C) or greater for a minimum of 3 days to meet the PFRP criterion. Additional operating requirements include a minimum active composting period of 21 days and a minimum curing period of 30 days. Monitoring of temperature and oxygen concentrations within the piles is required to document process performance. State approval of the design provided (1) that finished compost be monitored for agriculturally significant parameters (metals and nutrients), fecal coliforms, and certain chlorinated organic compounds; and (2) that records be maintained on the recipients of composted sludge. A buffer zone of 1,000 feet (ft) between the Hampton Roads facility and the nearest residence was also required as part of facility design, as were a runoff control system capable of collecting all runoff from a 10-year, 24-hour duration storm and an ambient air monitoring program.

The 1979 facilities plan projected sludge loadings of 8 dry tons per day (dtpd) at 20 percent total solids, or greater (40 wet tons per day (wtpd)), from each of the two wastewater treatment plants expected to contribute sludge to the composting facility. The Hampton Roads facility was ultimately constructed as an interim facility to process 10 dtpd (50 wtpd), with ultimate expansion to 16 dtpd (80 wtpd) if needed. Analysis of sludge from the James River treatment plant during facility planning indicated that finished compost metal content would not restrict product distribution.

Key features incorporated into the Hampton Roads facility during planning and design include the following: (1) use of wood chips as a bulking agent; (2) use of an uncovered concrete pad for mixing and screening operations; (3) all mixing and materials handling operations to be performed with front-end loaders; (4) active static pile composting on an uncovered concrete pad constructed with sunken aeration troughs to house flexible, perforated plastic aeration pipe; (5) extended static pile construction using a wood chip base and a blanket of finished compost; (6) negative aeration during active composting with odor control by filtering through a finished compost pile; (7) a covered concrete drying pad, open on all sides, with aeration piping located in the pad, to provide aeration of 3-ft-high windrows for 3 days to promote drying; (8) a minimum 30-day unaerated curing period; (9) uncovered storage of wood chips, unscreened compost, and finished compost in separate areas on unimproved soil; (10) unspecified screening equipment to allow for evaluation of various types; (11) gravel access roads; and (12) drainage of concrete pads to a central sewer for treatment at the James River treatment plant.

An evaluation of the market for finished compost was made as part of facility design. The evaluation included distributing questionnaires to determine the interest in and intended use of compost product and holding meetings with potential users. Target compost users included municipal and public works departments, golf courses, nurseries, hospitals and colleges, and private developers. Based on this evaluation, it was determined that a market for finished compost from the Hampton Roads facility could be developed.

Facility Characteristics

The Hampton Roads facility occupies a site having an operational area[2] of 6.2 acres (ac). A schematic of the site (not to scale) is shown on Figure 3. The current site includes 4.6 ac of concrete slabs and roadways. Storage areas, drying areas, and roadways comprise 3.0 ac. The active composting area is 1.1 ac, and the covered mixing and screening area is 0.5 ac. Covering is provided by a roofed structure, open on all sides. Additional information on site features is presented subsequently in Chapter 5.

The site has been modified since initial planning and design in response to operating problems which were experienced following construction. Unimproved wood chip and finished compost storage areas and gravel access roads have been paved to minimize wood chip losses, prevent finished compost contamination, and improve front-end loader maneuverability during wet weather. Also, to improve moisture control, mixing and screening have been relocated in the covered shed originally intended for drying. Drying operations have been moved to uncovered, concrete drying slabs, as shown on Figure 3.

Figure 4 shows the process schematic for current composting operations. Operation varies, depending on weather conditions. During dry-weather periods, composted sludge may be dry enough to allow direct screening, bypassing the separate drying step. However, since rainfall is fairly uniform throughout the year, separate-stage drying either before or after curing is the usual mode of operation.

Trucks equipped with power-ram horizontal-discharge trailers rated at 23 cu yd, transport dewatered sludge to the composting facility 5 days per week. Haul distance is 8 miles from the York River treatment plant, 6 miles from the James River treatment plant, and 35 miles from the Nansemond facility. Sludge deliveries are closely coordinated by facility operators to ensure effective and timely processing of incoming sludge.

[2]In this report, the term "operational area" is used to designate the area required for composting operations, including features such as buffer zones, access roads, and administrative facilities. The term excludes nonoperational area, such as that required for expansion, which may be available at the composting site.

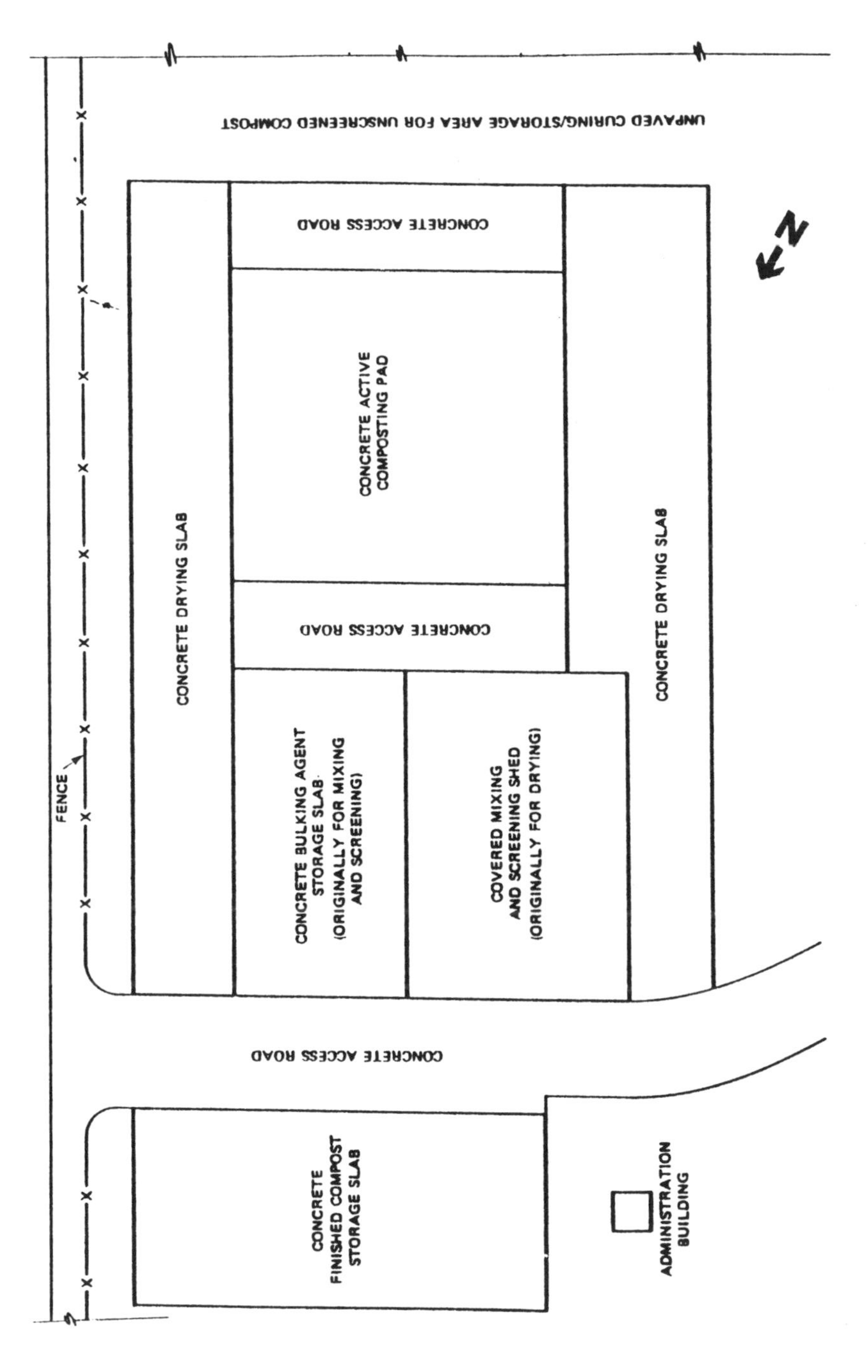

FIGURE 3. HAMPTON ROADS FACILITY SITE SCHEMATIC

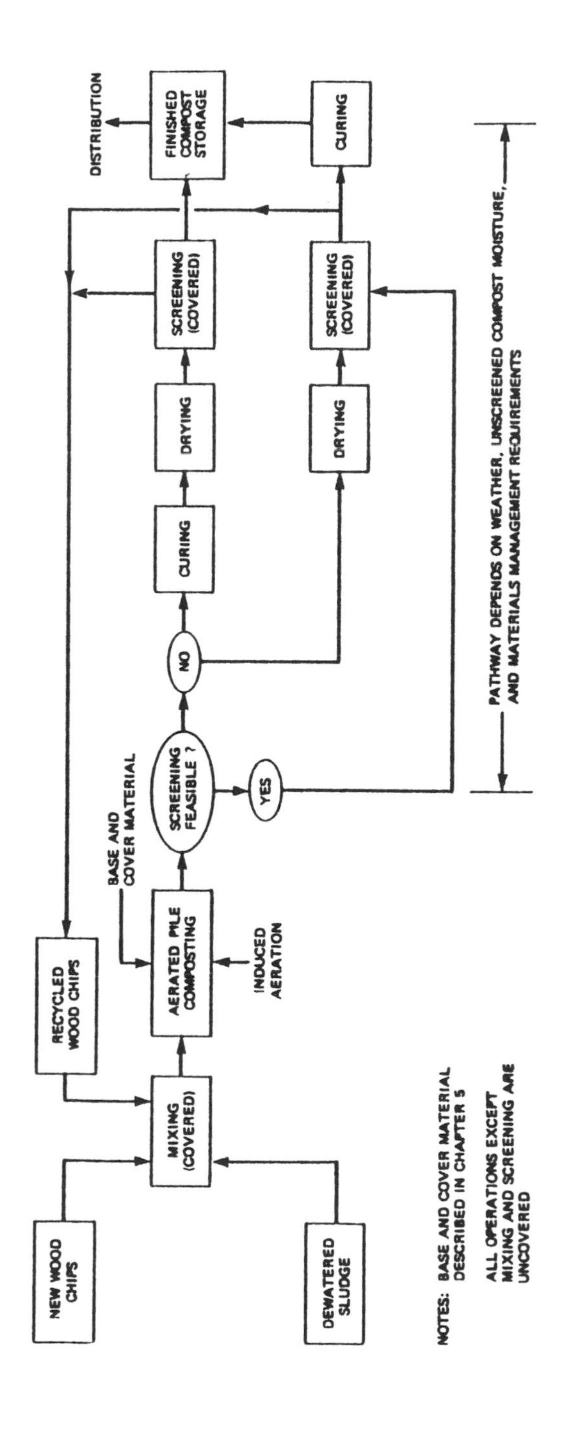

FIGURE 4. HAMPTON ROADS FACILITY PROCESS SCHEMATIC

Key equipment employed at the Hampton Roads facility is summarized below:

Equipment item	Number
Front-end loaders	
7 cu yd	1
5 cu yd	2
1.5 cu yd	1
0.75 cu yd	1
Power-ram semitrailers	2
Manure spreader, 18 cu yd	1
Utility tractors	2
Rototiller (tractor-pulled)	1
Pick-up truck	1
Mobile screens, vibrating deck	2

The 7- and 5-cu yd front-end loaders are used for various on-site operations, including mixing, materials transfer, pile construction, and similar operations. The loaders are equipped with roll-out buckets. The smaller front-end loaders are used for finished compost loading and for site housekeeping chores. The manure spreader is employed for fine mixing following front-end loader mixing of sludge and wood chips. Alternatively, the tractor-pulled rototiller is sometimes utilized for the fine mixing operation. The rototiller is also used to turn compost which is spread for drying. One mobile screen is located in the covered mixing and screening shed (Figure 3) as the primary screening equipment. The second is located near the unscreened compost curing and storage area and is used as a backup for the primary screen.

The compost facility is staffed by seven full-time operators, a chief operator (facility supervisor), and a part-time marketing agent. The seven full-time operators include four wastewater operators (highest level), two driver operators, and one operator-helper. Management and administration is provided by the James River plant manager and plant superintendent. The facility is operated 12 hours per day (two overlapping shifts) during the summertime and 8 hours per day during the wintertime, 6 days per week. Only pile teardown, screening, and cleanup is scheduled on the sixth day of each week.

SITE II FACILITY

The planning background, design considerations, and general features for the Site II facility are presented in this subsection.

Facility Planning and Design

The Site II facility utilizes the extended aerated static pile process to compost a mixture of raw (undigested) primary and secondary sludges containing lime and ferric chloride. Planning for the facility was initiated in the mid-1970s in response to a 1974 agreement between municipal jurisdictions contributing sewage to the Blue Plains Wastewater Treatment Plant located in Washington, D.C. This agreement specified, in part, that suburban jurisdictions would assist in disposal of raw municipal sewage sludge generated at the plant. Sludge generated at the Blue Plains plant is dewatered on vacuum filters using lime and ferric chloride for conditioning.

In 1977, six locations were selected by the Montgomery County Council for technical, environmental, and economic evaluation as candidate sites for a composting facility to process the County's share of raw sludge generated at the Blue Plains plant. The Washington Suburban Sanitary Commission accepted responsibility for developing the facility. After several years of intense public opposition and litigation to prevent construction that included concern for air and water quality protection, the Site II facility was opened in April 1983. Planning projections for the facility were made over a 30-year period, 1978 to 2008, with interim projections for 1983 conditions.

Selection of the aerated static pile process for application at Site II was based on experience at the demonstration facility of the U.S. Department of Agriculture's (USDA) Beltsville Agricultural Research Center and on a review of published literature. Alternative processes for composting the County's raw sludge were not considered.

The Site II facility was designed to meet State of Maryland regulations for composting municipal sludge. These regulations designate maximum allowable metals content based on USDA recommendations, and a minimum temperature of 55 degrees C for at least 3 days during the active composting period to meet the PFRP criterion. Distributed compost must be labelled as containing municipal sludge, and permits are required for application of compost on food crops or dairy cattle grazing land.

Compliance with Montgomery County, Maryland, regulations was also required as part of facility design. As a result, trucks used to transport sludge are fully enclosed and are capable of retaining all sludge in the event of a turnover. Sludge hauling is limited to the hours between 5 a.m. and 7 p.m. Site buffer zones of 1,000 ft for industries and 2,000 ft for residences were also required at the design stage.

The 1977 preliminary design was based on a 600-wtpd facility to treat raw sludge containing 18 to 20 percent solids with a 70 percent volatile solids content. An interim facility with a 400-wtpd capacity at 22 percent solids (88 dtpd) was designed and constructed.

Key features incorporated into the Site II facility during initial planning and design included the following: (1) use of wood chips as a bulking agent; (2) use of a paved, covered mixing area; (3) mixing of wood chips and dewatered sludge using mobile composters; (4) active static pile composting under cover; (5) pile construction using a wood chip base and a blanket of cured compost with aeration piping laid directly on a concrete composting pad; (6) an active composting period of 21 to 30 days with aeration rates up to 1,500 cubic feet per hour per dry ton (cfh/dt); (7) negative aeration during active composting, with a centralized exhaust gas collection system and odor absorption pile consisting of dry, screened compost and wood chips; (8) leachate and condensate collection and disposal; (9) optional covered, prescreen drying by spreading and rototilling; (10) fully enclosed screening operations using vibratory screens, with filtration of ventilated air for dust and odor control; (11) uncovered storage and curing of screened compost; (12) storage of wood chips, unscreened compost, and finished compost in separate areas on uncovered, concrete pads; and (13) separate collection and pond containment of runoff from process areas and from roof drains, with disposal at controlled rates to the local sewage collection system.

As part of facilities planning and design at the Site II facility, approval for unrestricted distribution of finished compost was obtained from the State of Maryland Department of Health. Approval required that finished compost be monitored every 2,000 cu yd for cadmium, lead, zinc, copper, nickel, and mercury.

Facility Characteristics

The Site II facility is situated on a 116-ac parcel of land; the operational area occupies about 40 ac, including containment ponds for runoff. The operational area exclusive of containment ponds is about 30 ac. A general layout for the facility is shown on Figure 5 and a process schematic for current operations is presented on Figure 6. The operational area exclusive of the containment ponds includes active composting pads, a new and recycled woodchip storage area, a mixing and screening building, and a finished compost storage and curing area. These areas, as well as the roadways, are paved. The active composting pads are covered and open on all sides. Two pads were provided at design, but one of these has been modified such that half of the pad is now used for aerated drying prior to screening (prescreen drying).

The sludge receiving and mixing area within the mixing and screening building is covered and open on two sides, while the screening area itself is completely enclosed. Other on-site features include odor control facilities, a general purpose building, administration and equipment maintenance buildings, access roads, weigh scale stations, a fuel dispensing area, and a laboratory. Key features of the process train are described in more detail subsequently in Chapter 5.

Completely enclosed, airtight trucks, required by Montgomery County ordinance and rated at 20 tons net capacity, haul dewatered sludge 32 miles from the Blue Plains treatment plant to the Site II facility. The haul route is approved by the state health department. Current sludge loadings require 18 to 20 truck loads per day and all incoming trucks are weighed upon arrival. Trucks are washed with a high pressure stream at the site and at the wastewater treatment plant.

A summary of key equipment used at the Site II facility is presented below:

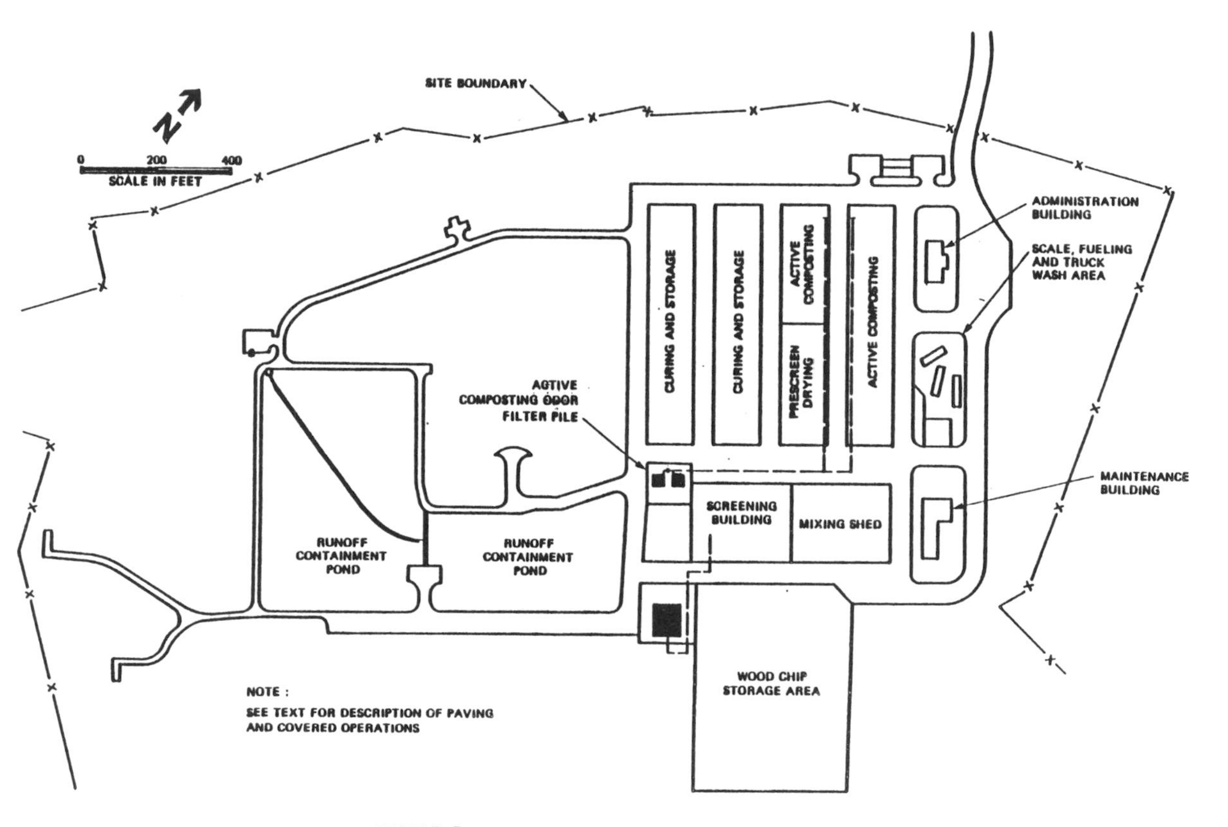

FIGURE 5. SITE II FACILITY SITE SCHEMATIC

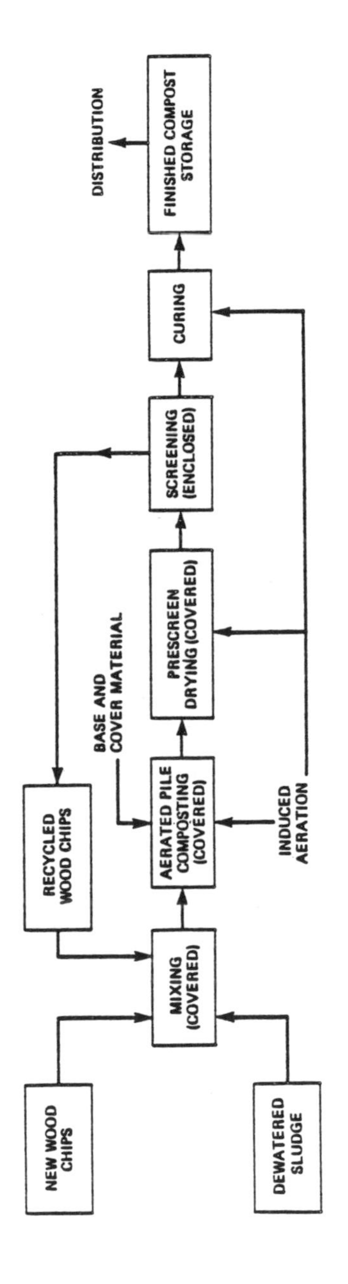

NOTE: BASE AND COVER MATERIAL DESCRIBED IN CHAPTER 6

FIGURE 6. SITE II FACILITY PROCESS SCHEMATIC

Equipment item	Number
Cobey rotoshredders	4
Front-end loaders	
10 cu yd	10
7 cu yd	2
Fixed, vibrating deck screens	3
Dump trucks	4
Pickup trucks, 1/2 ton	7
Flat-bed trucks, 1 1/2 ton	2
Utility tractors	2

The front-end loaders, which are equipped with roll-out buckets, are used for initial rough mixing of wood chips and dewatered sludge, which is followed by fine mixing by the mobile composters. Facility personnel report that two of the the mobile composters are sufficient for the mixing operation. Front-end loaders are also used to transfer the mix to dump trucks which then transport the material to the active composting area. Front-end loaders are again used to transfer the mix for pile construction and for pile teardown and prescreen drying restacking operations.

The Site II facility has a staff of 42, with functional categories as presented below:

Administration/supervision	5
Mixing operations	8
Composting operations	10
Screening operations	14
Laboratory	5
Total	42

The facility is in full operation with sludge deliveries and a complete staff 8 hours per day, 5 days per week. Because of screening problems, this activity is currently performed 16 hours per day.

COLUMBUS FACILITY

The planning background, design considerations, and general features for the Columbus facility are presented in this subsection.

Facility Planning and Design

The Columbus facility utilizes the extended aerated static pile process to compost a mixture of raw, unlimed primary and waste activated sludges. In April 1980, sludge processing facilities at the City's Southerly wastewater treatment plant were abandoned because of heat exchanger failure and resulting operation and maintenance difficulties, forcing the City to explore other means of processing sludge. In response to this emergency situation, design and construction of the composting facility was implemented. Design of the Columbus facility began on April 14, 1980, and the City awarded the construction contract on April 29, 1980. Construction started on May 12 and was essentially complete by July 22, 1980. However, construction of a portion of the facility was completed earlier, allowing the City to begin composting a small portion of sludge from the Southerly wastewater treatment plant by July 1. Equipment and materials were bid and were available at the time of start-up on July 1. Since the initial emergency design phase, facility design has been expanded and upgraded to include a mixing building, materials transport conveyors, a passive drying area, solar drying facilities, and administrative and maintenance facilities.

At the time of start-up, the State of Ohio had not established specific performance or operation criteria for composting. Because of the fast-track design/construction requirement, facility design was based on experience from the USDA's Beltsville, Maryland, aerated static pile demonstration facility. Other composting technologies were not evaluated. The Columbus facility was designed to meet a PFRP performance criterion of 55 degrees C in the coldest portion of a pile for 3 consecutive days. Additionally, a minimum 21-day composting period and a minimum 30-day curing period were provided at design.

Original design of the Columbus facility was based on processing an average quantity of 200 wtpd of dewatered raw sludge from the City's Southerly treatment plant. The maximum quantity anticipated at design was 300 wtpd. Solids handling operations at the Southerly treatment plant consist of primary and secondary sludge thickening followed by centrifuge dewatering of the combined raw sludge mixture. Based on dewatering performance at the time of composting facility design, a dewatered solids content of 20 percent was projected. The design sludge loadings are, therefore, equivalent to 40 dtpd, average, and 60 dtpd, maximum.

Analysis of raw solids from the Southerly treatment plant relative to product disposition was not performed during facility planning because of the emergency situation existing at the time of design. Composting was required in response to the emergency situation because land application of raw sludge is prohibited.

Key features of the Columbus facility as designed in 1980 included the following: (1) use of wood chips as a bulking agent; (2) an uncovered, concrete pad to be used for mixing, active composting and vehicle maneuvering and sloped to catch basins for drainage; (3) a two-step mixing operation to be performed using front-end loaders for rough mixing followed by fine mixing using a mobile composter; (4) pile construction using a wood chip base to cover aeration piping placed directly on the concrete pad and a daily cover of unscreened compost; (5) a 21-day active composting period using negative aeration based on a rate of 3 cubic feet per minute (cfm) per wet ton of sludge (900 cfh/dt); (6) use of an odor control filter pile composed of wood chips and unscreened compost, which was to be constructed for each daily extended pile compartment; (7) drying by positive aeration of the compost pile, in place, following the 21-day active compost period; (8) uncovered screening operations using equipment which was not specified at the time of design; (9) a minimum curing period (uncovered) of 30 days; (10) uncovered storage of wood chips and compost materials on graded fields adjacent to the concrete mixing and active composting pad; and (11) collection and pond containment of leachate, condensate, and storm runoff for transport to the Southerly wastewater treatment plant.

An evaluation of the market for finished compost was not made as part of facility design. Finished compost was to be stored on site until a market survey could be performed. Potential markets identified included agricultural applications, land reclamation, and sale as fuel for electric power generating plants.

Facility Characteristics

A layout showing current site features at the Columbus facility is presented on Figure 7. The figure includes modifications which have been made and are currently being implemented since initial design. Figure 8 shows the process schematic for current operations.

The concrete, active composting (and mixing) pad, and the leachate lagoon which were part of the original 1980 design are shown in the lower right corner of the site layout presented on Figure 7. In 1982, a curing and storage area west of the active composting pad was paved with concrete, and in 1983 a covered, enclosed mixing building was added. Screening facility improvements were also made in 1982. The leachate lagoon was enlarged in 1983. Solar drying, finished compost storage, and administrative facilities were added in 1984-85. This report does not include an evaluation of the solar drying facilities shown on Figure 7 because these were not operational at the time of the study. Current facilities are described in more detail subsequently in Chapter 5.

The area of the site is approximately 45 ac, including all the areas shown on Figure 7. The operational area exclusive of the solar facilities is about 37 ac.

Dewatered sludge is transported on a contract basis from the Southerly wastewater treatment plant to the Columbus facility in open-top, end-dump trailers of 25-ton capacity. Deliveries arrive between the hours of 5 a.m. and 2 p.m. Haul distance is about 10 miles. The treatment plant and compost facility managers determine when and how often sludge is to be transported to the composting site.

Key equipment available at the Columbus facility is summarized below:

Equipment item	Number
Cobey composters	2
Front-end loaders	
10 cu yd	2
7 cu yd	4
4 cu yd	2

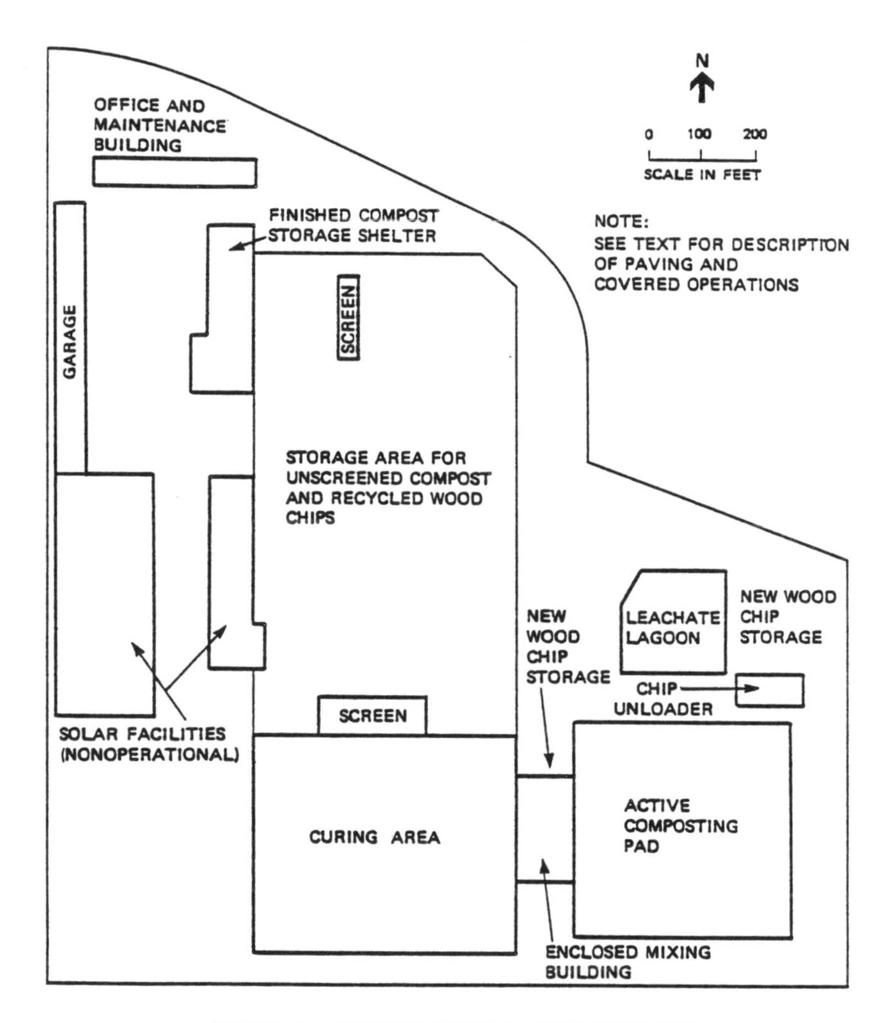

FIGURE 7. COLUMBUS FACILITY SITE SCHEMATIC

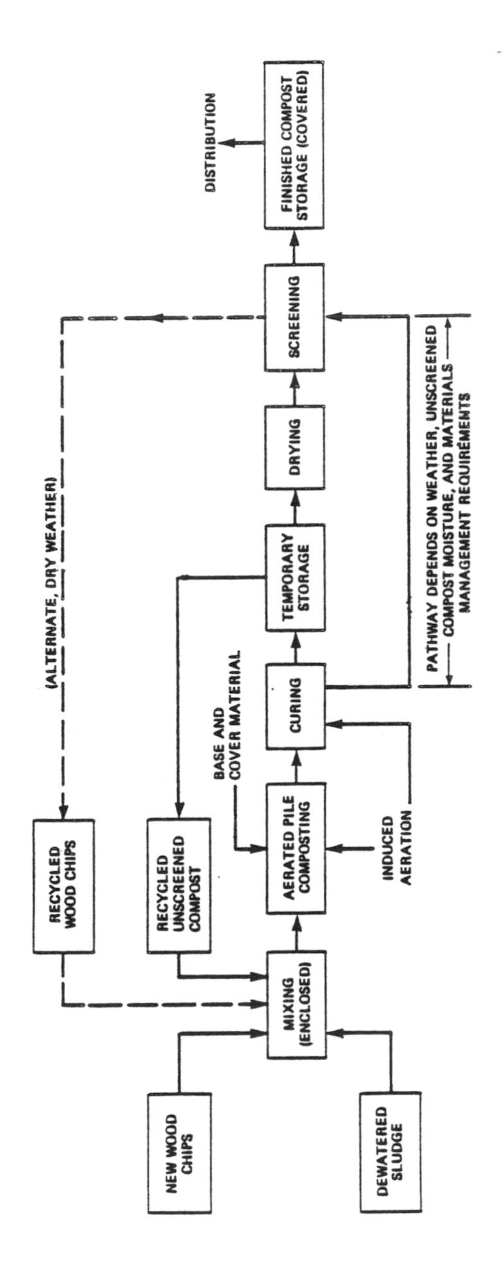

FIGURE 8. COLUMBUS FACILITY PROCESS SCHEMATIC

Equipment item (continued)	Number
Screens	
Vibratory, 3/8-inch openings	1
Rotary, 1/4-inch openings	1
Utility tractors with rototiller and harrow	2
Power-ram semitrailers	2
Trailer tippers	2
Street flusher truck	1
Pickup trucks and general purpose vehicles	3

A 60-ton scale is utilized to weigh incoming and outgoing material. There are two portable truck trailer tippers at the compost facility which are used for unloading wood chips delivered to the site. Each dumper is capable of safely tilting a fully loaded truck trailer to a minimum of 63 degrees from the horizontal in order to completely discharge its contents.

Two semitrailers are utilized for hauling composted materials or wood chips around the compost facility. Each unit consists of a truck tractor and a 28-cu yd, power-ram, horizontal-discharge trailer.

There are four front-end loaders with 7-cu yd buckets, two with 10-cu yd buckets and two with 4-cu yd buckets. The loaders are diesel powered, with four-wheel drive, and are articulated. The loaders transport, mix, stack, and load wood chips, sludge, compost mix, recycled wood chips, and compost. Two Cobey mobile composters are available for mixing at the Columbus facility. However, they are not utilized because of problems described in Chapter 5.

The Columbus facility has a staff of 19. Administrative personnel consist of the plant manager, a clerk/typist, and a laboratory technician. Operations personnel consist of 14 compost operators and two heavy-auto mechanics. The facility is operated 10.5 hours per day, 7 days per week, on a modified two-shift basis.

COMPARISON OF STATIC PILE SITE FEATURES

Key site features for each of the aerated static pile facilities investigated are summarized in Table 3. Operational areas for the Hampton Roads, Site II, and Columbus facilities are 6.2, 40, and 37 ac, respectively. As previously noted, operational area is defined as site area required for current composting operations, including access roads, buffer zones, and support facilities such as administrative buildings, maintenance structures, and truck scales. Storage ponds for runoff control, if required, are also included as part of the operational area. Land which is available on the site but is not a part of active operations, e.g., land available for expansion, is not included as part of the operational area. Operational areas employed at the Hampton Roads, Site II, and Columbus facilities are equivalent to about 0.12, 0.10, and 0.18 ac/wtpd, respectively, based on corresponding design sludge loadings of 50, 400, and 200 wtpd.

TABLE 3. KEY SITE FEATURES AT STATIC PILE FACILITIES

Site feature	Hampton Roads facility	Site II facility	Columbus facility
Operational area, ac[a]	5.2	40	37
Mixing operations			
Area, ac	0.3	1.1	0.4
Paved	Concrete	Concrete	Asphalt
Covered	Yes	Yes, enclosed two sides	Partially enclosed
Equipment	Front-end loaders with manure spreader or rototiller	Front-end loaders and mobile composter	Front-end loaders only
Bulking agent	Woodchips	Woodchips	Woodchips
Active composting			
Area, ac	1.1	2.5[b]	3.7
Paved	Concrete	Concrete	Concrete
Covered	No	Yes	No
Aeration piping	Laid in troughs in pad	Laid on pavement	Laid on pavement
Induced aeration	Positive and negative	Negative	Positive
Drying operations			
Area, ac	1.7	0.8[b]	-
Paved	Concrete	Concrete	-
Covered	No	Yes	-
Method	Spread and rototilled on open slabs	Restacking with induced aeration	-
Screening operations			
Area, ac	0.2	1.0	0.5
Paved	Concrete	Concrete	-
Covered	Yes	Fully enclosed	No
Equipment	Mobile screens	Fixed screens	Mobile and fixed screens
Curing operations			
Area, ac	-[c]	3.3[d]	4.6
Paved	No	Concrete	Concrete
Covered	No	No	No
Aerated	No	Yes	Yes
Storage operations			
Bulking agent			
Area, ac	0.4	4.5	0.9
Paved	Concrete	Asphalt	Concrete
Covered	No	No	No
Unscreened compost			
Area, ac	1.6[e]	-	10.0
Paved	No	-	Concrete
Covered	No	-	No
Finished compost			
Area, ac	0.4	-[f]	0.7
Paved	Concrete	-	Asphalt
Covered	No	-	Yes, partially enclosed
Runoff collection			
Ponds	No	Yes	Yes
Area, ac	-	10.0	0.9

[a]See text for site features included in operational areas presented.

[b]Prescreen drying area is one-half of original active composting pad. See text.

[c]Included as part of unscreened compost storage area.

[d]A 6-month storage/curing capacity for screened compost is provided.

[e]Includes curing. See footnote "c".

[f]Included as part of curing area. See footnote "d".

Key operations such as mixing are often covered to minimize moisture control problems, and at two of the three facilities the screening operation is also protected for this purpose. Most operating areas are paved either with concrete or asphalt, in part for runoff collection, in part for moisture control during the composting process, and in part for ease of operation during wet weather. Equipment and/or operating methods employed at various steps vary, as noted in Table 3.

Table 4 presents the unit area used currently for each operation, in square feet (sq ft) per wtpd, at the three static pile facilities investigated. The unit areas are based on design sludge loadings of 50, 400, and 200 wtpd at the Hampton Roads, Site II, and Columbus facilities, respectively.

Areas utilized for the static pile operations vary depending on various site-specific factors. For example, drying operations at the Hampton Roads facility are performed by spreading unscreened compost on uncovered, concrete slabs (see Figure 3), followed by rototilling. This operation employs 1,500 sq ft/wtpd. In contrast, intensive aerated prescreen drying currently employed at the Site II facility utilizes 100 sq ft/wtpd.

Another comparison shows that 10 ac (Table 3 and Figure 7) are provided for storage of unscreened compost at the Columbus facility because screening operations are affected by weather, and the market for distribution of finished compost is still being developed. This storage area requirement is equivalent to 2,200 sq ft/wtpd (0.05 ac/wtpd) and is one factor contributing to an overall unit operational area (0.18 ac/wtpd) which is considerably higher than the other static pile facilities investigated.

A third comparison shows that at the Site II facility, 1,100 sq ft/wtpd are associated with runoff containment ponds (see Figure 5). At the Columbus facility only 200 sq ft/wtpd are required, and at the Hampton Roads facility no area is employed for such facilities. If the area required for runoff containment ponds at the Site II facility is excluded, adjusted operational area for this facility is 30 ac. Adjusted unit area is about 0.08 ac/wtpd. There is little change in unit operational area at the Columbus facility if the runoff containment pond area is excluded.

TABLE 4. UNIT AREAS FOR STATIC PILE OPERATIONS

Operation	Unit area, sq ft per wet ton per day[a]		
	Hampton Roads	Site II	Columbus
Mixing	300	150	100
Active composting	950	300	800
Drying	1,500	100	-
Screening	200	150	100
Curing	-[b]	400[c]	1,000
Storage operations			
Bulking agent	350	500	200
Unscreened compost	1,400[d]	0	2,200
Finished compost	350	-[e]	150
Runoff collection	0	1,100	200

[a]Values rounded up to nearest 50 sq ft per wet ton per day.
[b]Included in unscreened compost storage area.
[c]Provides 6-month storage/curing of screened compost.
[d]Includes curing. See footnote "b."
[e]Included as part of curing area. See footnote "c."

4. Windrow Facility Descriptions

One conventional windrow facility, the Los Angeles County Sanitation Districts Joint Water Pollution Control Plant Composting Facility (Los Angeles facility) located in Carson, California, and one aerated windrow facility, the Metropolitan Denver Sewage Disposal District Number One Demonstration Composting Facility (Denver facility) located in Denver, Colorado, were investigated during the municipal sludge composting technology evaluation. A general description of each facility is presented in this chapter. Additional descriptive information is included in the windrow operations assessment presented in Chapter 6.

LOS ANGELES FACILITY

This subsection presents an historical perspective and current facility characteristics of the Los Angeles facility (1-9). This facility was not the product of extensive planning, evaluation, and design but, rather, has evolved over the years and has been the source of extensive research projects dealing with the fundamentals of the windrow composting process. Windrow composting technology as applied at the Los Angeles facility is continuing to evolve, due in part to a unique situation regarding product disposition and marketing which provides an incentive to experiment with methods for improving productivity.

Historical Perspective

By 1972, odor problems with static air drying of anaerobically digested sludge from the Joint Water Pollution Control Plant had become intolerable to the residential community that had built up around the wastewater treatment facility, and serious consideration was given to alternate methods of disposal, including landfill disposal and composting. Windrow composting experiments were conducted in the early 1970s using mobile composting machines to mix dried recycled sludge with fresh sludge cake. These experiments showed that when initial total solids content of the recycled and fresh sludge mixture reached 40 percent, and windrows were turned daily, aerobic conditions needed for composting were achieved. In addition, odor and fly problems associated with sludge drying decreased significantly.

At that time, all dewatered sludge (225 wet tons per day (wtpd)) from the Joint Water Pollution Control Plant was composted on a 10-acre (ac) unimproved dirt field. Elevated temperatures, over 60 degrees C, were being achieved within windrows that measured about 3 feet (ft) high by 12 ft wide at the base. Concerns related to pathogen destruction had not yet become a major issue. Limited monitoring data showed that Salmonella bacteria were being effectively destroyed. No monitoring for helminth ova or viruses was performed at this stage of composting development at the plant, and there were no odor complaints from neighbors regarding the composting operations.

In the early years of composting at the Los Angeles facility, a large tractor-pulled rototiller and a road grader were used to mix compost materials and build and turn the windrows. This method of windrow operation was very slow and led to pilot studies to evaluate different types of composting equipment. Two types of composting machines were evaluated in the mid-1970s: Terex-Cobey and Cobey Rotoshredder. The latter was selected for routine use.

In the late 1970s, fundamental research on the thermodynamic properties of composting systems led to an understanding of the relationships between energy available from the sludge cake and amendments and the energy requirements for composting and evaporation of water. There is an upper moisture content at which any composting system fails. This point is reached when the energy required for evaporation exceeds the available energy from the aerobic decomposition of organics present in the sludge and amendments. Theoretical calculations applied to the compost operation at the Los Angeles facility between 1972 and 1977 indicated that the facility operated at approximately 65 percent of the limit of thermodynamic failure (the point where the required energy just equals the available energy). Therefore, composting operations had a wide margin of safety, allowing considerable flexibility in carrying out the operations. To control the initial moisture content of the starting compost mixture at 40 percent total solids during this period, only 33 percent of the final compost product had to be recycled. The remainder of the finished compost was delivered to a private company, the Kellogg Supply Company, for independent sale and distribution.

Start-up of a new sludge dewatering facility at the Joint Water Pollution Control Plant using 44 basket centrifuges was planned for 1977. The basket centrifuges were designed to capture over 95 percent of the particles present in the plant's digested sludge, and as a result, the amount of dewatered sludge to be composted was projected to increase to over 1,600 wtpd. The projected post-1977 compost operation at the treatment plant involved handling a total of 4,750 wtpd (1,450 dtpd) of material, a seven-fold increase over pre-1977 levels.

In anticipation of the extra sludge quantities to be processed, the existing compost field was enlarged from 10 to 40 ac. Lime stabilization of soil in the compost field was employed. The upper 6 to 8 inches of soil were scarified, mixed with lime, and compacted. In addition, the following new equipment was purchased for the composting operation: (1) Flow-Boy semi-trailers to haul sludge to the field; (2) Athey force-feed loaders to remove dry compost from the field; and (3) specially designed, enclosed, end-dump trucks to be used in conjunction with the Athey loaders to remove finished compost from the field in a manner that minimized dust emissions.

Total solids content of dewatered sludge from the new basket centrifuges was projected to be 20 percent, which, when blended with 35 percent solids cake from existing centrifuges, would result in a blended sludge cake having a

25 percent total solids content. Applying thermodynamic principles to the post-1977 composting operation indicated that the new operation would be at 95 percent of the theoretical limit of thermodynamic failure. The post-1977 composting operation called for recycling 76 percent of the final compost product to maintain the initial total solids content of the windrow mixture above 40 percent (60 percent moisture content).

Start-up problems occurred when the new basket centrifuges were put in operation. Because of this, initial dewatered cake total solids from the new centrifuges were only 17 percent (at best) and frequently much less. This resulted in blended cake total solids as low as 19 percent (versus a projected value of 25 percent). As the new dewatering equipment was brought on-line, odors from the compost operation increased significantly and odor complaints began. When sludge cake production reached 900 wtpd, the frequency of odor complaints made it obvious that something must be done to immediately correct the problem. Thus, a decision was made to compost only that quantity of dewatered sludge that resulted in no odor complaints. The rest of the dewatered sludge was landfilled.

Eventually, start-up problems with the new basket centrifuges were solved by operating them in parallel with existing solid-bowl centrifuges. Although the original concept called for the basket centrifuges to operate in series by treating centrate from the solid-bowl machines, significant changes in sludge quality had occurred between the time the basket centrifuges were pilot tested in the early 1970s and the start-up of the full-scale system. This change was due principally to the implementation of an industrial user fee program requiring industries to pay a share of sewerage system operation and maintenance cost. As a result, many large paper companies located in the sewage collection area began to pretreat their wastes and recycle fiber. The reduction in paper fiber content of sludge to be dewatered had a significant effect on the performance of the centrifugal dewatering equipment. With the revised parallel operation, blended cake total solids content was increased to 23 percent. However, overall sludge particle capture was only 55 percent rather than the anticipated 95 percent. As a result, only 1,000 wtpd of dewatered cake could be produced rather than the 1,600 wtpd initially projected.

During the 6-year period from 1977 to 1983, experience at the Los Angeles facility showed that 500 wtpd is the maximum quantity of sludge that can be composted during the summer months without odor complaints. Due to lower productivities in the winter, the annual average is 425 wtpd. These values represent conversion of 6-day-per-week operation, as actually performed, to 7-day-per-week figures. Between 1977 and 1983, typical composting operations required that 45 percent of the compost product be recycled and mixed with fresh sludge cake to control moisture in the initial windrow mixture. No other amendments were used.

Various pieces of composting equipment were also replaced during the 6-year period from 1977 to 1983. period. The Athey loader was replaced by front-end loaders; the Flow-Boy trailers were replaced with end-dump and power-ram, horizontal-discharge trailers; and the Cobey composting machine then in use was replaced by a Scarab I mobile composter. Although the Cobey composter was used successfully for a number of years, it had originally been adapted from refuse composting operations, and facility personnel report that it appeared to have some limitations with denser material such as sewage sludge mixtures. Temperatures high enough to destroy pathogens were not being achieved, and the relatively high windrow surface-to-volume ratio favored heat loss. Also, windrow sizes and shapes were not uniform.

Four new anaerobic digesters were put into service at the Joint Water Pollution Control Plant in the fall of 1983. The new digesters improved solids stabilization by increasing solids detention time from 13 days to over 17 days. However, additional cellulosic material was degraded with the longer digestion time and, therefore, was no longer available as an energy source in the compost operation. From thermodynamic calculations, the composting operation at that time was operating just at the theoretical limit of thermodynamic failure. New low-speed scroll centrifuges were started up at this time, and sludge particle capture was increased to 95 percent. Sludge cake production increased to 1,500 wtpd. A new 25-ac paved compost field was also put into service. In 1984, the concept of using only dried, recycled compost for the bulking agent was abandoned, and the present practice of adding some sawdust or rice hulls to all starting mixtures was initiated.

Current Facility Characteristics

A layout of the current Los Angeles facility is shown on Figure 9. The site includes a 25-ac, asphalt-paved south composting field which is the main active windrow composting area currently in use. As previously described, the layout has been adapted since initial operations in the early 1970s to provide areas for equipment storage, research operations, and expanded composting operations. All operations are uncovered and only the south composting field is paved.

A breakdown of areas used for various operations at the Los Angeles facility is presented below:

Site feature	Approximate area, ac
South composting field	25
North composting field	10
Bulking agent storage	6
Finished compost storage	5.5
Equipment storage	0.5
Research area	3
Total site	50.0

Although total site area is about 50 ac, the principal areas routinely in use at the current loading of about 500 wtpd are the south composting area, bulking agent storage areas, equipment storage areas, and finished compost storage areas. Thus, the operational area for the Los Angeles facility is about 37 ac, which is equivalent to about 0.08 ac/wtpd.

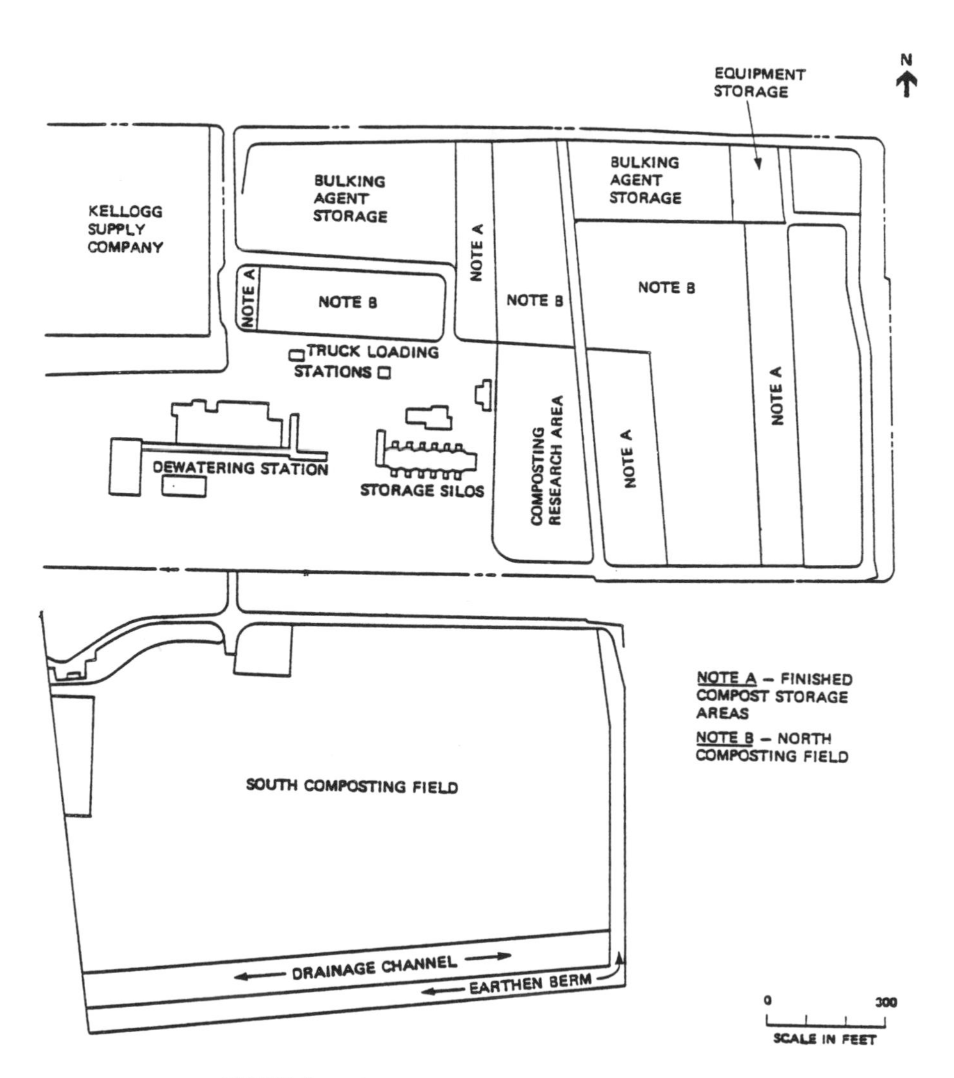

FIGURE 9. LOS ANGELES FACILITY SITE SCHEMATIC

Figure 10 shows a process schematic for current windrow composting at the Los Angeles facility. Several different amendments are used which yield four major compost products marketed by Kellogg Supply Company, as described below:

1. Nitrohumus--a general soil amendment product which is 90 percent composted sludge and 10 percent sawdust.

2. Topper--a top dressing product (new lawn covering, mulch) that is 60 percent sawdust and 40 percent composted sludge.

3. Amend--a product recommended for vegetable and flower gardens that is 75 percent rice hulls and 25 percent composted sludge.

4. Gromulch--an outdoor planting mix which is similar to Topper but contains other ingredients which Kellogg Supply Company considers proprietary.

Sludge from the Joint Water Pollution Control Plant dewatering building is conveyed to storage silos, and then from the silos to a truck loading station. These facilities are located adjacent to the composting area as shown on Figure 9. Twelve sludge storage silos have a capacity of 550 wet tons each, yielding a sludge total storage capacity of 6,600 wet tons. Sludge is stored in the silos during nighttime hours, over weekends, and during rainy periods when disposal (either composting or landfilling) is not possible.

Normal dry-weather operation calls for all silos to be empty by the end of the day shift on Fridays. By the end of the graveyard shift on Sundays, the silos are at their maximum normal dry-weather, in-storage levels as indicated below:

1. 3,000 wet tons (45 percent of capacity) for pre-1985 conditions.

2. 4,200 wet tons (64 percent of capacity) for 1985 conditions with addition of 200 million gallons per day (mgd) of pure oxygen secondary treatment at the treatment plant.

From Monday to Friday, storage levels oscillate downward, reaching zero by Friday afternoon. One operator at a central control panel can control sludge withdrawal from all 12 silos, as well as the operation of conveyor belts leading from the silos to two truck loading stations (see Figure 9). Each truck loading station requires one operator.

Trucks which haul sludge from the storage silos to the composting field are loaded with approximately 15 tons of wet cake and then moved to another loading area where amendment (rice hulls, sawdust, and finished compost) is added from a stockpile. Sludge is measured using weigh bins, whereas amendment is measured by the bucket volume of a front-end loader. Loaded trucks are then driven to the compost pad where windrows are formed. All trailers have a 42-cu yd capacity, and both end-dump and power-ram, horizontal-discharge trailers are used.

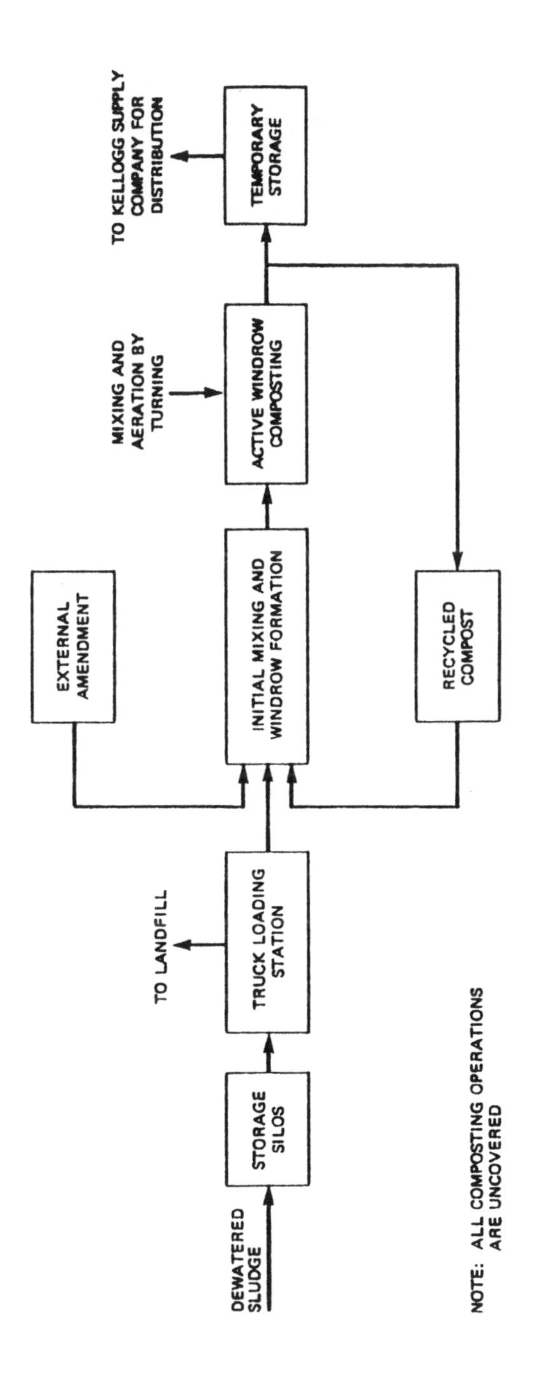

FIGURE 10. LOS ANGELES FACILITY PROCESS SCHEMATIC

Key equipment employed at the Los Angeles facility is summarized below:

Equipment item	Number
Front-end loaders	2
Scarab composters	
7 tons per minute rating	3
11 tons per minute rating	1
Semitrailers, 42 cu yd	3
Water truck	1

Use of this equipment is described in Chapter 6.

The Los Angeles facility is staffed by ten full-time compost operators plus one foreman. Management and administration are provided by staff at the Joint Water Pollution Control Plant. Compost operators are not trained wastewater treatment plant operators but, rather, truck drivers. The composting facility is operated 10 hours per day, Monday through Friday, and 8 hours per day on Saturday.

DENVER FACILITY

The planning background, design considerations, and general features of the Denver facility are presented in this subsection.

Planning Background

The Denver facility was developed as a demonstration facility utilizing modified windrow methods with induced aeration (aerated windrow) to compost a mixture of anaerobically digested primary and waste activated sludges. Conventional windrow methods were also used for comparison to the aerated windrow technique. Planning for the demonstration facility was initiated in 1982 as an outgrowth of comprehensive sludge management planning for Metropolitan Denver Sewage Disposal District Number One's Central Wastewater Treatment Plant and the City of Denver's Northside Wastewater Treatment Plant (10).

Prior to 1981, disposal of sludge from the Central and Northside plants was by land disposal at a remote site. Opposition to continued disposal at this site led to a value engineering study in 1981, which analyzed over 30 possible sludge management alternatives and recommended that windrow composting of dewatered, digested sludge be evaluated as one of several final alternatives for long-term (through 2004) sludge management (11). Both a remote site and a site at the Central treatment plant were to be considered for the composting facility location. After composting, the compost product would be distributed in accordance with applicable regulations.

Subsequent evaluation of capital and operation and maintenance costs for each alternative showed that composting at either a central or remote site would be the least costly long-term method for sludge management. Following this evaluation, a study to demonstrate the feasibility of windrow composting utilizing a 10-dtpd facility was initiated. The intent of this demonstration study was to provide information from which a full-scale composting facility could be designed and constructed.

Design Considerations

The demonstration study of windrow composting was performed as part of planning for a dual-utilization, long-term sludge management arrangement for handling sludge from the Central and Northside wastewater treatment plants. Under this arrangement, sludge would be land-applied at a remote location whenever weather permitted, and composting would be used as a backup when weather conditions prohibited agricultural land application.

The Denver facility was designed and the demonstration study performed to evaluate site-specific factors for implementing composting under the dual-utilization plan. Because of this, results presented in this report must be viewed within this context, rather than from the perspective of a full-scale, operating facility. Demonstration facility design was initiated in June 1982 and construction was initiated in October 1982.

Digested sludge dewatering at the Central treatment plant is accomplished by centrifugation with polymer addition. A dewatered sludge total solids content of 16 percent is routinely achieved, even with variable thickened sludge feed. Dewatered solids volatile content is typically about 60 percent. Under the dual-utilization sludge management plan, dewatered solids content of digested sludge for composting was projected to be in the range of 13.5 to 18 percent. Centrifuge considerations limit dewatering capability to about 18 to 20 percent total solids. These upper levels require lower feed rates, lower scroll speeds, and increased polymer dosages. Because of this, total solids content of dewatered sludge transferred to the composting facility is typically maintained at about 16 percent, rather than at the higher values which are achievable with the existing dewatering equipment.

Draft State of Colorado regulations governing land application of domestic sewage sludge establish four grade classifications, based on the maximum allowable concentrations of five metals (cadmium, copper, lead, nickel, and zinc) and polychlorinated biphenyls. To be classified in a particular grade, the finished sludge must meet all concentration limits. Grade 1 sludges are those having the most stringent concentration limits.

The draft regulations also state the following: (1) that all municipal sewage sludge be stabilized prior to land application and beneficial use; (2) that Grade 1 sludge, if dewatered, may be applied to any land for any beneficial use if it has first been subjected to a process to further reduce pathogens (PFRP), or has been stored for 1 year; (3) that Grade 2 sludges and those Grade 1 sludges not meeting the requirements identified in (2) can only be applied to agricultural or disturbed lands; and (4) that maximum allowable application rates and cumulative loadings on beneficial use sites not be exceeded. Composting may be utilized either for sludge stabilization (a process to significantly reduce pathogens (PSRP)), or as a PFRP under the draft regulations.

Facility Characteristics

As shown on Figure 11, the site for the Denver facility is located adjacent to the Central wastewater treatment plant. The Denver facility consists of (1) approximately 6 ac of asphalt pavement for conventional and aerated windrow composting, as well as mixing and curing; (2) an induced aeration system; and (3) a storm runoff, leachate, and condensate collection system. A portion of the paved composting area is covered to study the effect of covering on windrow composting performance. Figure 12 is a schematic of the covered structure which is open on all sides. The aeration system was designed such that either positive or negative aeration of windrows can be used. Piping arrangements allow exhaust air quality (odor) evaluation during suction aeration. All of the water (storm runoff, leachate, and condensate) from the site is collected and returned to the Central plant. Trucks equipped with power-ram horizontal-discharge trailers transport dewatered sludge from the adjacent Central plant to the composting facility.

Under the dual-utilization sludge management plan, the Denver facility was designed to study composting to provide a dry, stabilized product for ultimate distribution. Objectives and key features of the demonstration study performed at the Denver facility are described in more detail in Chapter 6 of this report. The demonstration facility design was oriented to investigate the application of the aerated windrow process at this location. A schematic of the aerated windrow process used at the Denver facility is shown on Figure 13.

As described in Chapter 1, the aerated windrow process is similar to the conventional windrow process with the exception that facilities for induced aeration are provided.[3] Incoming dewatered sludge from the Central treatment plant is mixed with an amendment such as recycled compost[4] and formed into windrows. The windrows are turned periodically using a mobile composter over a 30- to 45-day active composting/drying period. Turning renews the surface area (providing aeration) and ensures that all material has adequate residence time within the pile interior. Induced aeration is also provided during this period in the aerated windrow process. After the 30- to 45-day active composting period, finished compost is either recycled as an amendment for new windrow development or transferred for product distribution.

Four key features were incorporated into the aerated windrow demonstration study at Denver (10). First, based on a review of experience at other composting locations a minimum initial mix total solids content of

---

[3]In addition to aeration provided by periodic turning with mobile composters.

[4]Other amendments and amendment mixtures were investigated as part of the demonstration study. These are as described in a subsequent section of this chapter.

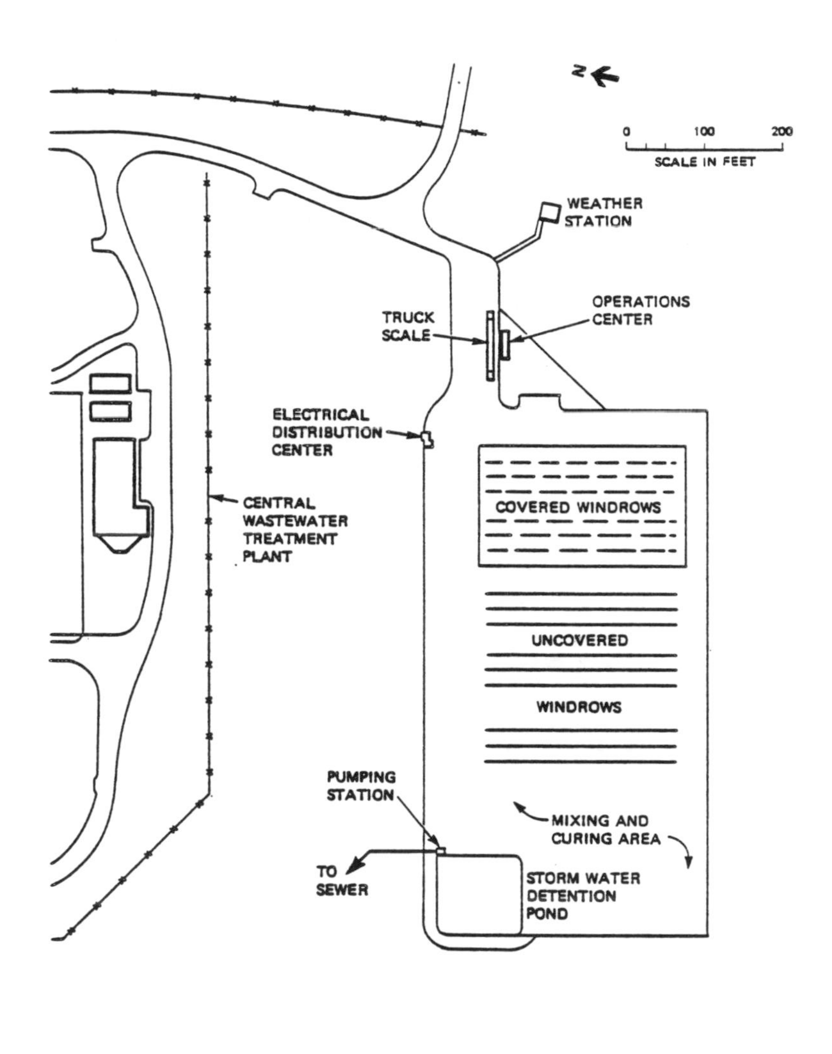

FIGURE 11. DENVER FACILITY SITE SCHEMATIC (ADAPTED FROM REFERENCE 10)

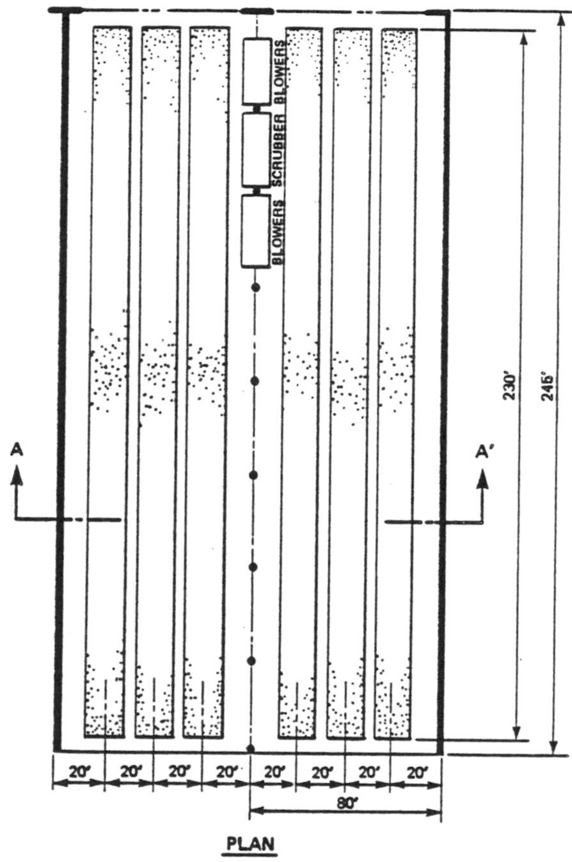

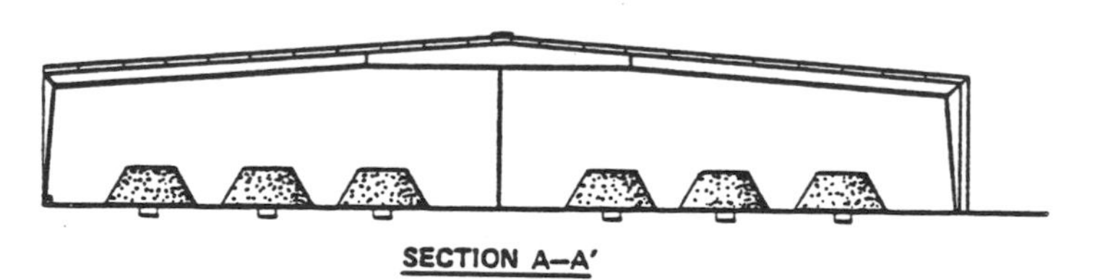

FIGURE 12. COVERED WINDROW STRUCTURE AT DENVER FACILITY (ADAPTED FROM REFERENCE 10)

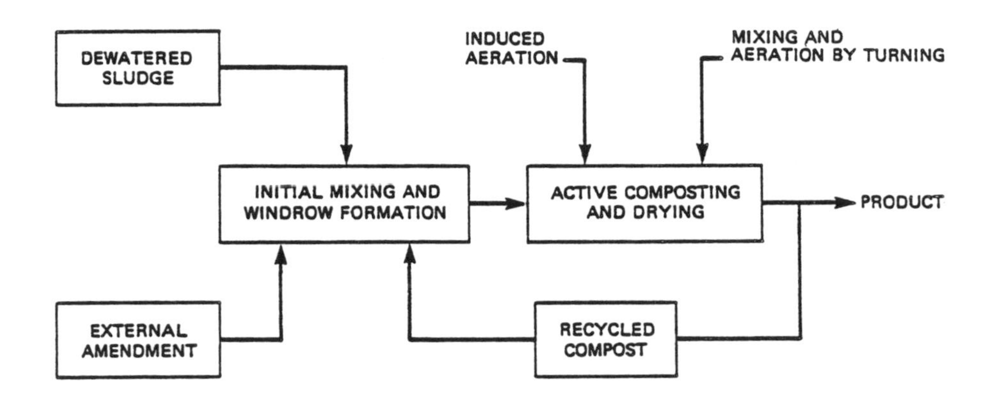

FIGURE 13. DENVER FACILITY AERATED WINDROW PROCESS SCHEMATIC

40 percent was considered a necessary criterion to provide sufficient porosity for effective aeration and successful composting. Secondly, it was determined that if only recycled compost were used as an amendment, the windrow process would have to achieve a final total solids content of 60 percent to maintain cost-effective materials handling, machinery needs, and space requirements. Thirdly, an induced aeration system was incorporated into facility design to evaluate the effectiveness of this technique for enhancing the composting process, reducing odor, and, in particular, to assess drying capabilities. Finally, covered and uncovered windrow areas were provided (see Figure 11) to evaluate the impact on performance of covering the windrow process. These features are discussed in more detail in Chapter 6.

Key mobile process equipment utilized at the Denver facility are summarized below:

Equipment item	Number
Brown Bear	1
Scarab mobile composter	1
Front-end loader	1

The Brown Bear is a Model ABR 12T and is used for windrow formation. The Scarab mobile composter is rated at 5 tons per minute, has a tunnel 5 ft high by 14 ft wide, and is used for mixing and aeration during active composting. The front-end loader is equipped with a 2-1/2-cu yd bucket and air conditioned cab.

External Amendment Characteristics

A variety of amendments other than recycled compost were investigated during the demonstration study. Physical characteristics of the external amendment materials employed are presented in Table 5. Average values, ranges, and the number of observations on which averages are based are included in the table.

REFERENCES

1. Caballero, R. "Experience at a Windrow Composting Facility: Los Angeles County Site. In Sludge Composting and Improved Incinerator Performance, EPA Technology Transfer. July 1984.

2. Hay, J.C., et al. "Disinfection of Sewage Sludge by Windrow Composting." National Science Foundation Workshop on Disinfection, Coral Gables, Florida. May 7-9, 1984.

3. Iacoboni, M.D., et al. "Windrow and Static Pile Composting of Municipal Sewage Sludges." EPA, Municipal Environmental Research Laboratory, Cincinnati, Ohio. May 1982.

4. Garrison, W.E. "Composting and Sludge Disposal Operations at the Joint Water Pollution Control Plant." County Sanitation Districts of Los Angeles County, Whittier, California. June 23, 1983.

TABLE 5. PHYSICAL CHARACTERISTICS OF EXTERNAL AMENDMENTS FOR DENVER STUDY

		Solids concentration, percent				Bulk density, pounds per cubic foot		
		Total		Total volatile				
Amendment	Number of observations	Mean	Range	Mean	Range	Number of observations	Mean	Range
Straw	6	82.2	68.1-90.8	83.0	73.2-89.8	4	56.6	25-111
Cornstalks	1	91.1	-	91.5	-	-	-	-
Tree trimmings	7	69.8	62-86.4	75.8	50.1-84.3	3	27.1	20.7-38
Sawdust	15	64.8	48.2-90.9	85.2	55.2-99.6	5	33.7	21.4-45
Wood chips	7	73.5	59.8-82.2	87.1	82.9-99.2	2	33.5	29.8-37
Bark chips	1	60.3	-	98.7	-	1	29.5	-

5. LeBrun, T.J., et al. "Overview of Compost Research Conducted by the Los Angeles County Sanitation Districts." National Conference on Municipal and Industrial Sludge Composting, Philadelphia, Pennsylvania. November 17-19, 1980.

6. Hay, J.C., et al. "Forced-Aerated Windrow Composting of Sewage Sludge." Virginia Water Pollution Control Association Conference, Williamsburg, Virginia. April 30-May 2, 1984.

7. Iacoboni, M. "Compost Economics in California." Biocycle, July-August 1983.

8. Iacaboni, M., et al. "Deep Windrow Composting of Dewatered Sewage Sludge." National Conference on Municipal and Industrial Sludge Composting, Philadelphia, Pennsylvania. November 17-19, 1980.

9. Horvath, R.W. "Operating and Design Criteria for Windrow Composting of Sludge." National Conference on Design of Municipal Sludge Compost Facilities, Chicago, Illinois. August 29-31, 1978.

10. Black and Veatch/Engineers-Architects. Central Plant Facility Plan, Volume IV. Prepared for the Metropolitan Denver Sewage Disposal District Number 1. October 1983.

11. Culp, Wesner, and Culp, Consulting Engineers. Value Engineering Study. Prepared for the Metropolitan-Denver Sewage Disposal District Number 1. February 23-27, 1981. As reported in Reference 10.

5. Static Pile Composting Operations Assessment

This chapter presents an assessment of current operations at three aerated static pile composting facilities investigated during the technology evaluation: (1) the Hampton Roads Sanitation District Peninsula Composting Facility (Hampton Roads facility) located in Newport News, Virginia; (2) the Washington Suburban Sanitary Commission Montgomery County Composting Facility (Site II facility) located in Silver Spring, Maryland; and (3) the City of Columbus, Ohio, Southwesterly Composting Facility (Columbus facility) located in Franklin County, Ohio. The chapter is organized in three main subsections. The first subsection describes the operations assessment methods, the second presents the results of the assessment, and the third provides a postconstruction comparison of the facilities studied.

OPERATIONS ASSESSMENT METHODS

The static pile operations assessment is based on on-site investigations at each of the three static pile composting facilities. The dates of the on-site investigations are presented below:

Facility	Dates of on-site investigations
Hampton Roads	June 26 to 29, 1984 July 16 to 19, 1984 August 20 to 23, 1984
Site II	November 26 to 29, 1984
Columbus	December 17 to 20, 1984 January 7 to 12, 1985 February 11 to 15, 1985

The on-site investigations consisted of a review of plant records; on-site observations of facility operations; and, at the Hampton Roads and Columbus facilities, independent process testing. Plant records which were reviewed included, where available, (1) facility planning and design documents, (2) operation and maintenance (O&M) manuals, (3) process monitoring and materials management documents, (4) equipment maintenance records, (5) administrative documents, and (6) reports of special studies. On-site observations included interviews with plant personnel, time and motion studies, and general observations of facility activities (e.g., housekeeping). Independent process testing was performed at the Hampton Roads and Columbus facilities, but not at the Site II facility.

Information was sought from each aerated static pile facility in six major categories related to design and operation: (1) sludge loadings and characteristics, (2) sludge transport, (3) process technology, (4) physical design, (5) environmental controls, and (6) product disposition. These categories were selected to provide a means of comparing the facilities investigated and to serve as a basis for identifying generic features of static pile composting. Unique and/or site-specific features of each facility were also documented.

Key information related to sludge loadings and characteristics which was sought at each facility included variability in sludge moisture and volatile content, average and peak sludge loadings, heavy metal and nutrient content of sludge, and factors related to treatment plant and sludge dewatering operations. Sludge transport features sought included mode, haul distance, delivery frequency, reliability, and equipment applications.

Process technology features were of particular interest. Bulking agent and/or amendment use, mixing methods and effectiveness, and pile configuration characteristics were documented as they relate to process operation and performance. Aeration requirements and temperature and oxygen monitoring and control requirements were also investigated. Information on screening effectiveness was obtained, as well as on drying and/or curing steps. Routine plant operating records and independent measurements were utilized to establish representative materials balances for an installation processing anaerobically digested sludge (the Hampton Roads facility) and one processing raw sludge (the Columbus facility).

Equipment applications and operating and maintenance requirements were documented where feasible to do so using plant records. Time and motion studies were performed at two of the installations.

Physical design features investigated included the following: (1) effectiveness of site layout from an operating perspective, (2) general materials management methods, (3) staffing, and (4) maintenance considerations. Actual capital and O&M cost information was obtained at each facility from which unit costs were developed.[5] Costs associated with facility improvements were also documented. On-site O&M costs associated with bulking agent or amendment use, utilities, personnel, aeration pipe replacement, equipment, and similar activities were sought, as were O&M costs associated with sludge transport, marketing programs, and general administration. Procedures applied to control odor, dust, weather, leachate, condensate, runoff, and <u>Aspergillus fumigatus aerospora</u> were investigated, where feasible to do so.

[5]Cost evaluations are presented in Chapter 7.

Finished compost quality relative to dryness, uniformity, heavy metal content, nutrients, and biological characteristics was documented from routine plant monitoring data and also by independent measurements at two facilities. Storage methods employed to protect the quality of finished compost, to meet fluctuating demands for product, and to otherwise meet materials management requirements were investigated, as were means of product transport where this function is part of facility operations.

OPERATIONS ASSESSMENT RESULTS

The results of the static pile operations assessment are presented in this subsection. The findings are organized according to the following categories: (1) sludge loadings and characteristics; (2) precomposting operations (mixing and pile construction); (3) active static pile composting, including aeration and process control requirements; (4) postcomposting operations (pile teardown, drying, curing, screening, and finished compost quality and distribution); and (5) general facility operations, including materials flow considerations and environmental control features. For each category of the static pile operations assessment, site-specific information obtained during the technology evaluation is initially presented and discussed. At the end of the text for each category, general comments and observations from the assessment are presented.

Sludge Loadings and Characteristics

Representative sludge loadings and dewatered sludge total solids contents for the three aerated static pile facilities investigated are presented in Table 6. General sludge characteristics are presented in Table 7. Current loadings vary from about 65 to 360 wet tons per day (wtpd) on an operating-day basis, which translate to 11 to 62 dry tons per day (dtpd) for a dewatered sludge total solids content at each plant, as received, of 17 percent. On a calendar-day basis, sludge loadings at the Hampton Roads, Site II, and Columbus facilities are 54 wtpd (9 dtpd), 260 wtpd (44 dtpd), and 170 wtpd (29 dtpd), respectively. Peak-to-average day loading ratios vary from 1.4 to 1.9. Although the mean sludge total solids content at each facility is about 17 percent, variations as low as 13 to 14 percent and as high as 22 percent are sometimes received.

Volatile content of the anaerobically digested sludge processed at the Hampton Roads facility is typically about 55 percent, with a range of 43 to 61 percent. Raw, unlimed sludge received at the Columbus facility is typically 74 percent volatile, with a range of 63 to 79 percent. Liming of raw sludge to a pH of 12.0 to 12.5 (Table 7) prior to composting results in volatile solids contents typically in the range of 34 to 51 percent at the Site II facility.

Slludge characteristics presented in Table 7 are provided primarily for reference purposes. Except as noted in the table, data reported are from monitoring records at each of the facilities. Independent testing during the technology evaluation was performed only at the Hampton Roads and Columbus facilities, and, with two exceptions, confirmed plant monitoring results for

TABLE 6. REPRESENTATIVE SLUDGE LOADINGS FOR STATIC PILE FACILITIES

Item	Hampton Roads facility[a]	Site II facility[b]	Columbus facility[c]
Type of sludge processed	Anaerobically digested primary and secondary solids	Raw, limed primary and secondary solids	Raw, unlimed primary and secondary solids
Loadings per operating day[d]			
Average, wet tons	63	360	170
Average, dry tons	11	62	29
Peak-to-average ratio	1.6 to 1.8	1.4 to 1.6	1.9
Operating days per week	6	5	7
Total solids, percent			
Mean	17	17	17
Range	14 to 22	15 to 22	13 to 22
Total volatile solids, percent			
Mean	55	43	74
Range	43 to 61	34 to 51	63 to 79

[a]Based on operating records for January to June 1984.

[b]Based on operating records for January, July, and September 1984.

[c]Based on operating records for January to May 1984.

[d]An operating day is any day on which process-related activities, such as pile construction, pile teardown, screening, or similar operations are performed.

TABLE 7. SLUDGE CHARACTERISTICS FOR STATIC PILE FACILITIES

Constituent[a]	Concentration range[b]		
	Hampton Roads facility	Site II facility	Columbus facility
Bulk density, lb/cu yd	(1,600)	-	(1,900)
Trace minerals, mg/kg			
Cadmium	10-12	<0.2-6	7-30
Chromium	72-124	-	180-350
Copper	650-751	143-294	200-300
Lead	192-237	95-295	190-340
Mercury	(1.8-2.7)	1.2-2.9	-
Nickel	44-56	<2-36	55-170
Zinc	1,210-1,776	238-652	1,400-2,200
Nitrogen, as N, percent			
Total Kjeldahl	(5.0-5.8)	0.4-4.8	-
Ammonia	-	0.1-0.3	-
Total phosphorus, as P, percent	(2.6-3.1)	0.06-2.2	-
General minerals, mg/kg			
Calcium	18,250-31,000	-	-
Chlorides	-	4,000-9,000	-
Iron	-	40,260-79,934	-
Magnesium	2,500-5,500	-	-
Potassium	1,175-1,375	747-974	-
Sodium	-	53-622	-
Sulfates	-	500-3,080	-
Soluble salts, as $CaCO_3$, percent	-	3.9-5.9	-
pH	(8.2-8.3)	12.0-12.5	-

[a]Dry weight basis, except density and pH

[b]Based on plant monitoring data, except parenthesized results are from indepedent testing during the on-site investigations.

these facilities. Sludge from the James River treatment plant of the Hampton Roads facility yielded higher concentrations of calcium (32,500 milligrams per kilogram [mg/kg] dry weight) and magnesium (9,100 mg/kg), based on averages of three samples, than previously reported (18,250 and 5,500 mg/kg). The cause of the difference was not determined during the technology evaluation.

Independent testing at the Columbus facility yielded higher concentrations of cadmium and zinc (38 and 2,580 mg/kg, respectively) than values reported in Table 7. These results are based on averages of three sludge samples collected during the on-site investigations. Again, however, the cause of the difference was not determined during the technology evaluation.

Wastewater treatment plant operations which affect the character of digested and/or dewatered sludge can carry over to the composting facility. For example, the Hampton Roads facility currently receives dewatered sludge from three wastewater treatment plants (see Chapter 3)--York River, James River, and Nansemond. The James River plant typically achieves 2.5 percent total solids after digestion, with dewatered solids of 15 to 17 percent. During the on-site investigations, personnel at this plant indicated that dewatered solids in this range are produced when digested solids fed to the belt filters are in the range of 2.4 to 2.6 percent total solids and polymer is applied at 20 to 21 pounds per dry ton, whereas dewatered solids content drops to about 15 percent if digested sludge solids content drops to 2.3 percent, even at higher polymer doses. Under worst-case conditions, dewatered solids of 13 to 14 percent are produced at this plant.

As a second example, the Site II facility processes raw (undigested) sludge from Washington, D.C.'s Blue Plains wastewater treatment plant which is chemically treated with lime and ferric chloride prior to dewatering using vacuum filters. Dewatered sludge is stored in bins prior to transport to the composting facility. Although the dewatering operation is capable of producing dewatered sludge having a total solids content of 22 percent, the solids are difficult to remove from the storage bins at this concentration, and, therefore, water is added prior to transport. This typically results in a sludge total solids content, as received, of about 17 percent.

Sludge loadings received at composting facilities can be highly variable, as illustrated by the data presented in Table 8. This table summarizes sludge loading characteristics at the Site II facility for 3 months--January, July, and September 1984. Total quantity of sludge processed in 1 month varied from 5,859 wet tons in September to 9,573 wet tons in July, and the total number of piles constructed varied from 36 to 56 (September and July, respectively). On an operating-day basis, average sludge loadings varied from 307 wtpd in January to 456 wtpd in July. On a calendar-day basis, average loadings varied from 195 wtpd in September to 309 wtpd in July. Peak-day sludge loadings were 471, 664, and 514 wtpd in January, July, and September, respectively, yielding corresponding peak-to-average loading ratios of 1.53, 1.46, and 1.58 based on average operating-day loadings.

The Columbus facility provides another example of variable sludge loadings. This facility received an average of 170 wet tons of sludge per day for the period from January 1, 1984, through May 28, 1984 (Table 6).

TABLE 8. SITE II SLUDGE LOADING CHARACTERISTICS

Item	Month, 1984		
	January	July	September
Quantity processed, wet tons	6,760	9,573	5,859
Number of days per month			
Operating	22	21	18
Calendar	31	31	30
Pile construction			
Total number	48	56	36
Maximum number in one day	3	4	3
Maximum sludge quantity in one pile, wet tons	197	224	200
Sludge loadings, wet tons			
Average per operating day	307	456	325
Average per calendar day	218	309	195
Peak day	471	664	514

However, daily loadings varied from a low of 39 tons per day to a peak of 326 tons per day. The peak-to-average day ratio for this period was 1.9. Note that the Columbus facility operates 7 days per week, and thus loadings on an operating-day basis are equivalent to those on a calendar-day basis. Variability in sludge loading is an important factor to consider for design since operating area, materials handling, bulking agent quantity, manpower utilization, and other process requirements are impacted by the daily quantity of sludge received.

A comparison of actual and design sludge loadings for each static pile facility investigated is presented in Table 9. The comparison shows that the average total solids content of sludge being received at each facility is below that estimated at design by 3 to 5 percent. Current sludge loadings on a wet ton per operating day basis are 126, 90, and 85 percent of design loadings at the Hampton Roads, Site II, and Columbus facilities, respectively. However, because actual sludge total solids are below those projected at design, corresponding current sludge loadings on a dry ton per operating-day basis are 110, 70, and 73 percent of the respective design values.

Current peak-to-average-day loading ratios for the Site II facility (1.4 to 1.6) are similar to the value projected at design (1.5). However, the ratio at Columbus is higher (1.9 versus 1.5). A peak-to-average-day loading ratio was not projected at design for the Hampton Roads facility; however, current values (1.6 to 1.8) are higher than the design value of 1.5 which was used at the other two facilities.

Higher-than-anticipated peak-to-average-day loading ratios can cause several operational problems, including (1) extra time spent for mixing and pile construction, (2) a concomitant reduction in personnel time available for other operations, and (3) faster-than-expected utilization of active composting area. Generally, high peak loadings occur for only a limited duration and do not cause major operational impacts for more than several days at a time.

Estimates of sludge volatile solids used for facility design were not readily available during the technology evaluation. Representative current values for each facility which are presented in Table 6 are typical of anaerobically digested sludges, heavily limed raw sludge, and unlimed raw sludge.

Based on the assessment of sludge loadings at the Site II, Hampton Roads, and Columbus facilities, two key factors for effective composting were identified, as described below:

1. Sludge loadings established for design should consider variability in both the quantity and quality of sludge to be processed. Although loadings based on calendar days are useful for general comparisons, operating-day loadings more accurately reflect day-to-day materials management and process requirements. For example, when translating annual sludge generation rates from per capita suspended solids

TABLE 9. ACTUAL AND DESIGN SLUDGE LOADINGS FOR STATIC PILE FACILITIES

Item	Hampton Roads facility	Site II facility	Columbus facility
Design estimates			
Sludge total solids, percent	20	22	20
Sludge loadings			
Average wet tons per day	50[a]	400[b]	200[a]
Average dry tons per day	10	88	40
Peak-to-average day	Not specified	1.5	1.5
Current operations			
Sludge total solids, percent			
Average	17	17	17
Minimum	14	15	13
Sludge loadings per operating day			
Average wet tons	63	360	170
Average dry tons	11	62	29
Peak-to-average day	1.6 to 1.8	1.4 to 1.6	1.9
Sludge loadings, percent of design			
Wet ton basis	126	90	85
Dry ton basis	110	70	73

[a]Operating or calendar-day basis not defined.

[b]Considered nominal capacity. Design considers operating days but does not specifically designate the loading per operating day.

criteria to sludge loading values, 5-day-, 6-day-, or 7-day-per-week operation should be defined. Operating flexibility to respond to day-to-day variability in sludge total solids (moisture) content, as well as the wet quantity of sludge received, also needs to be considered. Bulking agent use, mixing operations, personnel utilization, equipment utilization and effectiveness, area requirements for storage and process functions, and composting performance are some of the features which are impacted by loading variations.

2. The potential impact of wastewater treatment plant operations on loading variability should also be considered during the design of a municipal sludge composting facility. In addition to technical evaluations such as those needed to define the effectiveness of dewatering operations, establishing management procedures for coordinating sludge deliveries can minimize the potential for odor generation, particularly with raw sludge and/or during hot weather, and provides for efficient materials handling during composting.

Precomposting Operations

Precomposting operations include both mixing and pile construction activities. An assessment of precomposting operations at the three static pile facilities investigated is presented below.

Mixing Operations--

Wood chips are used as a bulking agent at all of the aerated static pile facilities investigated. Both hardwood and softwood chips are employed, though personnel at one facility prefer hardwood chips because they do not decompose as rapidly as softwood chips and are less expensive to purchase. Chips are generally stored in piles on paved surfaces without any cover. Paved surfaces prevent wood chip losses and help to minimize moisture control problems during inclement weather.

Dewatered sludge is mixed with recycled and new wood chips at volumetric ratios of about 3.5 to 4.5 cubic yards (cu yd) of wood chips per cu yd of sludge at the sludge total solids concentrations typically received at the composting facilities studied. The mix ratio varies with chip quality, moisture content of sludge and chips, season, and the proportion of fresh and recycled chips used. All facilities employ mobile equipment for the mixing operation. A minimum initial total solids concentration of 40 percent in the wood chip-sludge mixture is considered essential for effective static pile composting. Mixing operations performed at each of the static pile facilities are described in detail below.

At the Hampton Roads facility, mixing is performed on a 0.3-ac portion of concrete pad located under a roofed structure, open on all sides (see Figure 3). Dewatered sludge is mixed with recycled and new wood chips to produce a mixture containing a minimum of 40 percent total solids. The mix ratio of wood chips to sludge varies from 3.5:1 during the summer to 4.0:1 for wintertime operation. Based on a dewatered solids content of 20 percent projected at design, the expected mix ratio under design conditions was estimated at 2.5 to 3. Thus, current operation at a representative solids concentration of about 17 percent (Table 6) requires a 30 to 40 percent increase in wood chip use over the original design estimate.

The higher volumetric mix ratio utilized at the Hampton Roads facility during wintertime operation has been established from operating experience for moisture control in the initial mix. It also helps to achieve an adequate moisture content during the drying phase, which enables the screening operation to operate efficiently. Visual characterization of dewatered sludge is used by operators to adjust the wood chip volume for each load of sludge to maintain the minimum total solids content of the mix to be composted. Actual measurement of wood chip volume, sludge volume, and moisture are also used for process control.

During routine operation, trucks equipped with power-ram trailers discharge dewatered sludge onto a bed of wood chips or, in some instances, directly onto the concrete mixing slab. Each truck delivers about 23 cu yd of sludge which is then mixed with the required amount of wood chips, about 80 to 90 cu yd depending on season. Thus, total volume of the initial mix varies from about 100 to 115 cu yd. Based on 11 observations taken on two separate days during the on-site investigation at this facility, unloading time (including truck cleanup) averaged 10 minutes, and average total on-site time of each truck was 23 minutes.

Currently, the Hampton Roads facility employs a two-step mixing operation consisting of coarse mixing of sludge and wood chips by a front-end loader followed by fine mixing using either a manure spreader or a tractor-mounted rototiller. Originally, only front-end loader mixing was utilized, but numerous 3-inch- to 12-inch-diameter sludge balls remained in the mix. The two-step method requires the same total mixing time (35 to 45 minutes) and produces a more uniform mixture. Typically, coarse mixing is accomplished with a 7-cu yd front-end loader equipped with a roll-out bucket and requires 20 to 25 minutes. The mixture is then deposited into an 18-cu yd manure spreader or spread out to a 12-inch depth for rototilling.

Various mixtures at the Hampton Roads facility were independently tested for pore space and homogeneity as part of the on-site investigation at this location. The rototilling method provided the most uniform mixture, as presented below:

Mixing method	Pore space, percent	Clods, percent
Front-end loader only	44	5 to 20
Front-end loader/manure spreader	60	10
Front-end loader/rototiller	62	5

For increased process control during wet weather, the mixing operation at the Hampton Roads facility has been relocated under a covered shed originally intended for aerated drying (Chapter 3). The covered shed is open on all sides, and rainwater can enter the mixing area during heavy storms. This does not generally affect initial mix moisture control but is inconvenient for operating personnel and can affect materials handling efficiency.

At the Site II facility, new and recycled wood chips used as a bulking agent are stored in an uncovered, 4.5-ac storage area adjacent to a partially enclosed mixing building. The storage area is paved and drained for collection of surface runoff. Wood chips are obtained from land clearing operations, and because chip quality is not controlled according to design specifications, occasionally composting operations are impaired. Wood chips are sampled periodically for moisture content.

Mixing is performed on a 250-ft by 200-ft (1.1-ac) covered concrete pad which is enclosed on two sides. Dewatered raw sludge is mixed with recycled and new wood chips to produce a mixture containing a minimum of 40 percent total solids. The ratio of wood chips to sludge typically varies from 4 to 4.5 cu yd per wet ton (cu yd/wt), depending on the moisture content of the sludge and wood chips and the ratio of new to recycled wood chips. Loads requiring higher blending ratios are occasionally received. When this occurs, they are rejected and returned to the wastewater treatment plant. Typically, the volume of recycled chips used for mixing is 60 to 100 percent of the total wood chip volume required for mixing.

The design mix ratio for the Site II facility was 2.5 cu yd/wt; thus, current wood chip use is 60 to 80 percent greater than that projected at design. The large increase in wood chip use has increased wood chip storage and material handling requirements.

A mixing foreman at the Site II facility controls the mixing operation by visual inspection. Routine measurements of sludge and wood chip volumes, and moisture contents of each mix component are also obtained for process control. Data recorded for each load include sludge tonnage, volume of old chips used, volume of new chips used, and mix ratio. Every third sludge truck is sampled for pH, moisture, total solids, and volatile solids.

During routine operations at the Site II facility, sludge is dumped on a 40-cu yd bed of new wood chips which has been placed to a depth of 12 inches by a front-end loader. Initially, material is premixed and shaped into a rough windrow with the front-end loader, followed by fine-mixing using a mobile composter. Based on experience, facility personnel report that three passes with the mixer are required to provide a uniform mix. The windrow is visually inspected after the third pass, and if sludge clumps are present, additional passes are made until the desired uniformity is achieved.

At the Columbus facility, an enclosed mixing building was designed to control initial mix moisture content during inclement weather and to reduce odor. The building is 90 ft wide, 175 ft long, and 32 ft high. There are two 40-ft-wide by 30-ft-high doors on each end, and five 20-ft-wide by 16-ft-high doors on each side of the building; all doors are electrically operated. There are no support columns within the building, and the interior clear height is 30 ft, which allows space for end-dump trucks to unload. The building is provided with an asphalt floor and an odor control ventilating system.

Personnel at the Columbus facility report several problems with the mixing building. Size is insufficient, and, thus, doors are opened during mixing to provide adequate space for maneuvering front-end loaders. This negates the effectiveness of the odor control system. The floor has buckled in several locations and this affects front-end loader maneuverability. Fine mixing with mobile composters (as designed) is not performed because floor discontinuities prevent their use. Fog sometimes occurs inside the building limiting visibility; there have also been problems with the doors and there have been drainage problems.

During routine operations at the Columbus facility, two compost operators are on site when the first sludge truck arrives at approximately 5 a.m. They mix sludge with new wood chips and recycled wood chips or unscreened compost and stack the mixture as required for static pile composting. Typically, facility operators are able to construct two beds in preparation for sludge deliveries. Fresh wood chips used in the mixing operations are not normally analyzed for moisture content, and the contract for wood chips does not specify a maximum moisture content. Front-end loaders mix the sludge and wood chip mixture.

The mixing operation at the Columbus facility is performed by different operators, and mixing time is varied. Thus, to assess mixing effectiveness at this facility, a time and motion study of mixing operations was performed during the on-site investigations. For fine-mixing operations, mixing time varied from 5 to 40 minutes. Although measurements to determine mixing effectiveness were not obtained, the 40-minute mix appeared uniform, with good porosity and few clumps of unmixed material. The 5-minute mix, on the other hand, was not homogeneous and had numerous sludge balls and clumps of wood chips. This assessment indicates that having more than one person responsible for mixing impacts the effectiveness of the mixing operation and therefore should be avoided. Facility personnel report that this procedure has been necessary because of split shifts and because parts of the site have been under construction for the last 5 years, making it difficult to develop a routine. The procedure is reportedly to be rectified when solar drying facilities recently constructed at the Columbus facility go on-line (see Chapter 3).

In summary, the assessment of mixing operations at the three static pile facilities investigated during the technology evaluation indicates that mixing of sludge and wood chips to an initial total solids concentration of at least 40 percent (60 percent moisture) is an important criterion for effective composting. To achieve a uniform blend of sludge and wood chips using the types of mobile mixing equipment and mixing techniques employed at the facilities investigated, a mixing time of about 40 to 45 minutes is required. Based on operating problems experienced at the Hampton Roads and Columbus facilities, paved wood chip storage and mixing areas enhance moisture control and minimize wood chip losses during inclement weather. The need to cover wood chip storage areas does not appear to be critical. However, for the geographic regions in which the facilities are located, covered or enclosed mixing is important. Finally, bulking agent use at the facilities studied is greater than that projected at design by as much as 80 percent.

Pile Construction--

Static pile construction techniques vary depending on site-specific conditions. However, generally one of two basic schemes is employed. One configuration involves placing perforated aeration piping of an appropriate length directly on a paved compost pad, after which the piping is covered with a base material, such as wood chips, 12 to 18 inches deep. Sludge-wood chip mixture is then placed on the base material in an extended pile configuration (see Chapter 1, Figure 2), after which an insulating cover material is applied. The aeration piping is connected by a manifold to a blower which provides either positive or negative aeration. Generally, one such extended pile compartment is constructed for each daily loading of dewatered sludge.

The second basic static pile configuration is similar to the first except that perforated aeration piping is placed in belowground troughs which are formed as part of the paved compost pad. The troughs are then filled with a material such as wood chips, after which base material, sludge-wood chip mixture and cover material are placed in a manner similar to the first configuration. The aeration piping is again connected to a blower via a manifold.

Although many aerated static pile composting facilities have been constructed with a single blower to service more than one extended pile compartment, the trend is now toward providing one blower of sufficient capacity for each daily extended pile compartment constructed. One blower may be provided to service the entire length of a pile compartment, or one may be provided at each end to service one-half of each pile compartment.

Key pile construction features at each static pile facility studied are presented in Table 10 and on Figures 14 through 18. Extensive testing by plant personnel was required at two of the facilities to arrive at reliable pile construction techniques for routine operation, as described below.

At the Site II facility, the composting area consists of two adjacent covered concrete pads, 120 ft by 600 ft, which have a minimum clear working height of 20 ft.[6] A roadway approximately 42 ft wide is provided between the two composting pads for servicing aeration equipment. The concrete pads are sloped to drain leachate and condensate to a sewer. The Site II facility constructs daily extended pile compartments which are about 110 ft long by 20 ft wide by 12 ft high (Figure 14). Extended pile compartment construction entails four key steps: (1) placing four 6-inch-diameter aeration pipes at 5-ft intervals on the compost pad, as shown on Figure 15; (2) placing a 12-inch bed of base material over the piping system; (3) placing the sludge-

[6]Use of one of these pads has been modified since initial construction such that half is used for active composting and half is used for aerated drying prior to screening (see Chapter 3, Figure 5).

TABLE 10. KEY STATIC PILE CONSTRUCTION FEATURES

Item[a]	Hampton Roads facility	Site II facility	Columbus facility
Pile dimensions, daily compartment, ft			
Length	110	110	200
Width	8-10	20	8
Height	13	12	12-13
Base material			
Depth, in	12-18	12	12
Central pile	New wood chips	New wood chips	New wood chips
Toes (distance from pile end)	Unscreened compost (10 ft)	Unscreened compost (12 ft)	Same as central area
Cover material			
Depth, in	18 (summer) 24 (winter)	18 -	18 -
Type	Unscreened compost	Screened or unscreened compost	Unscreened compost
Special cover	4-6 in of screened compost on face of toes	-	-
Aerating piping			
Placement	Laid in trough cast into composting pad	Laid directly on paved composting pad	Laid directly on paved composting pad
Type	Rigid	Flexible	Flexible
Material	PVC	Plastic	Polyethylene
Diameter, in	6	6	5
Pipe laterals			
Number per daily compartment	1	4	2
Lateral length unperforated, ft	22	15	16
Lateral length perforated, ft	88	86	184
Perforation spacing	Uniform	Variable	Uniform
Perforation area, sq in per linear ft	0.88	0.415-4.28	4.00
Aeration blowers[b]			
Total number	8	42	40
Capacity, each, hp	3	15	1
Capacity, total, hp	24	630	40
Unit capacity, hp per wtpd[c]	0.48	1.58	0.20

[a]Values tabulated are typical for each facility.
[b]Active composting only.
[c]Based on design loadings of 50, 400 and 200 wtpd for the Hampton Roads, Site II and Columbus facilities, respectively.

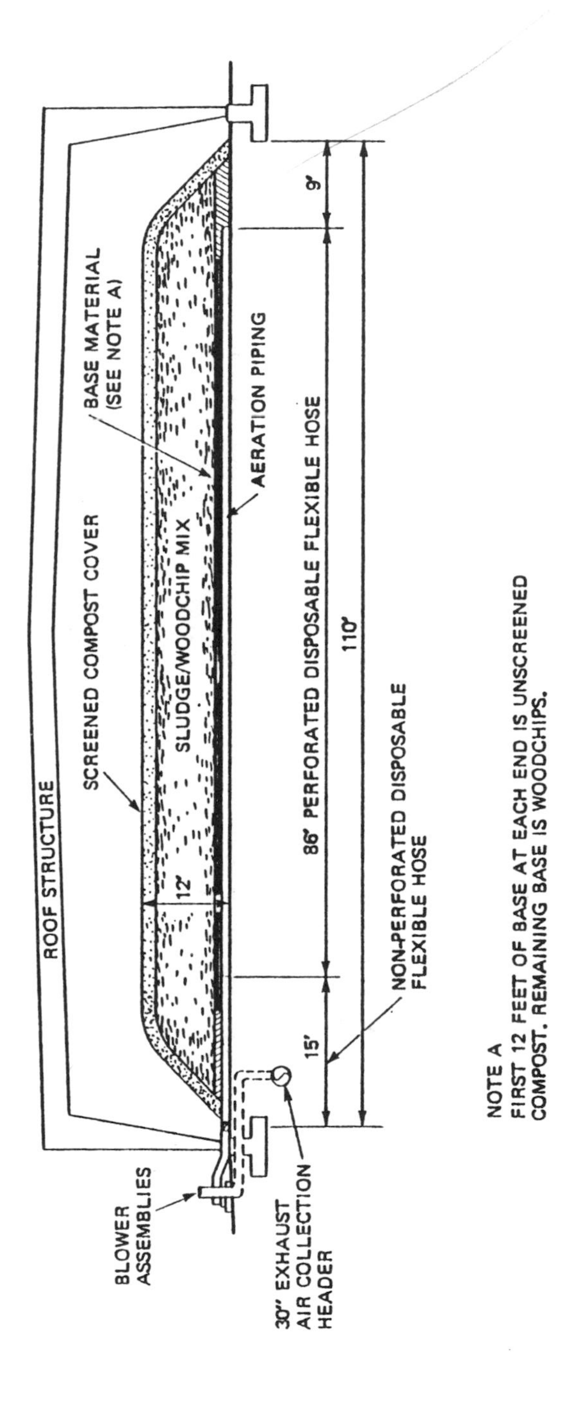

FIGURE 14. PILE CONSTRUCTION AT SITE II FACILITY

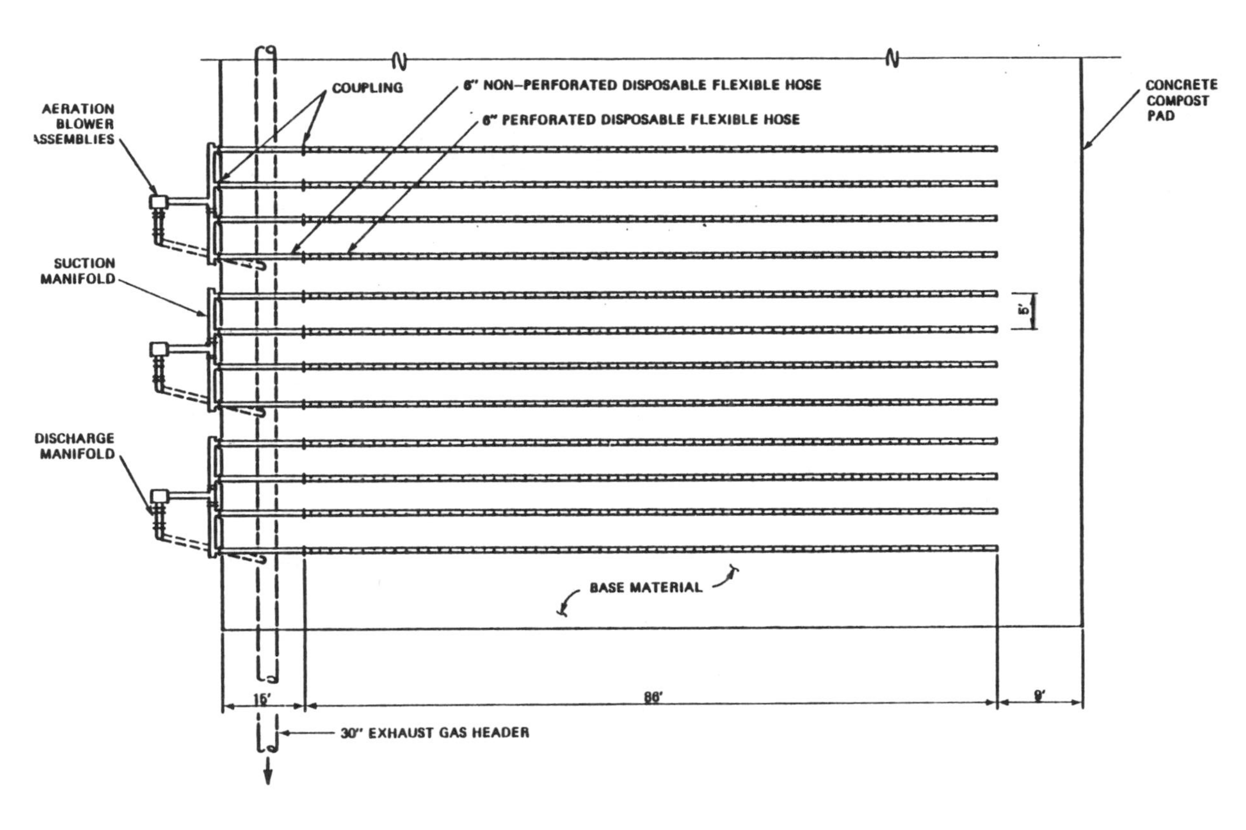

FIGURE 15. SITE II FACILITY AERATION SYSTEM CONFIGURATION

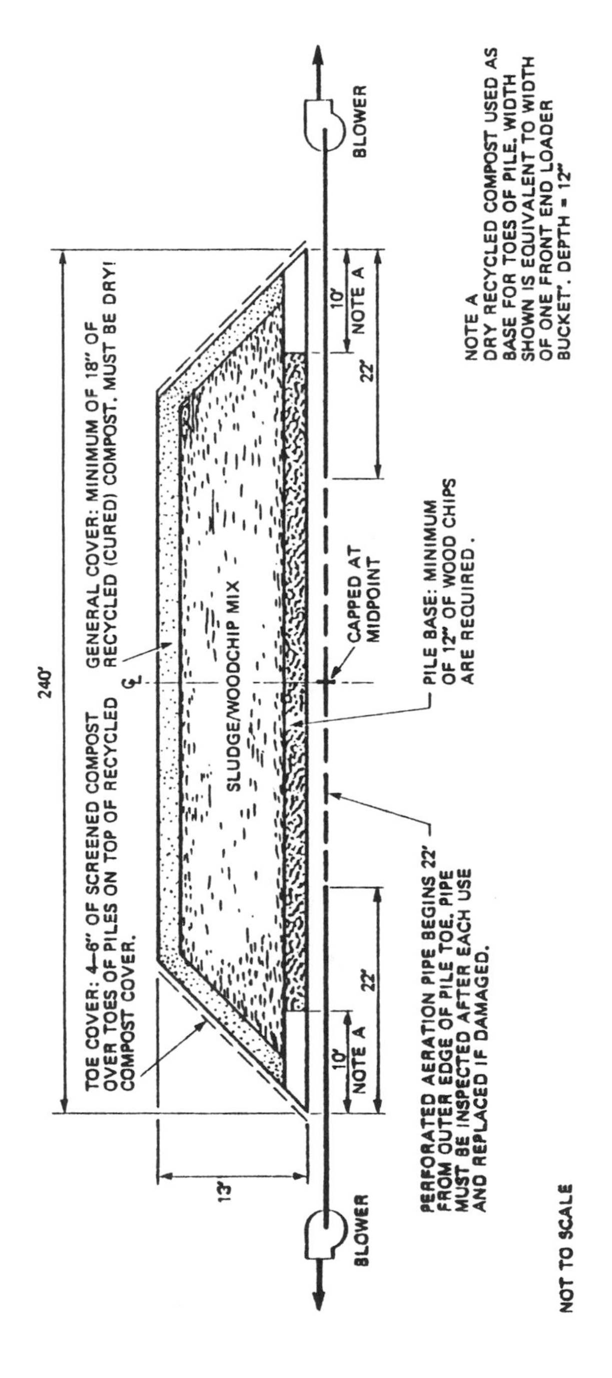

FIGURE 16. PILE CONSTRUCTION AT HAMPTON ROADS FACILITY

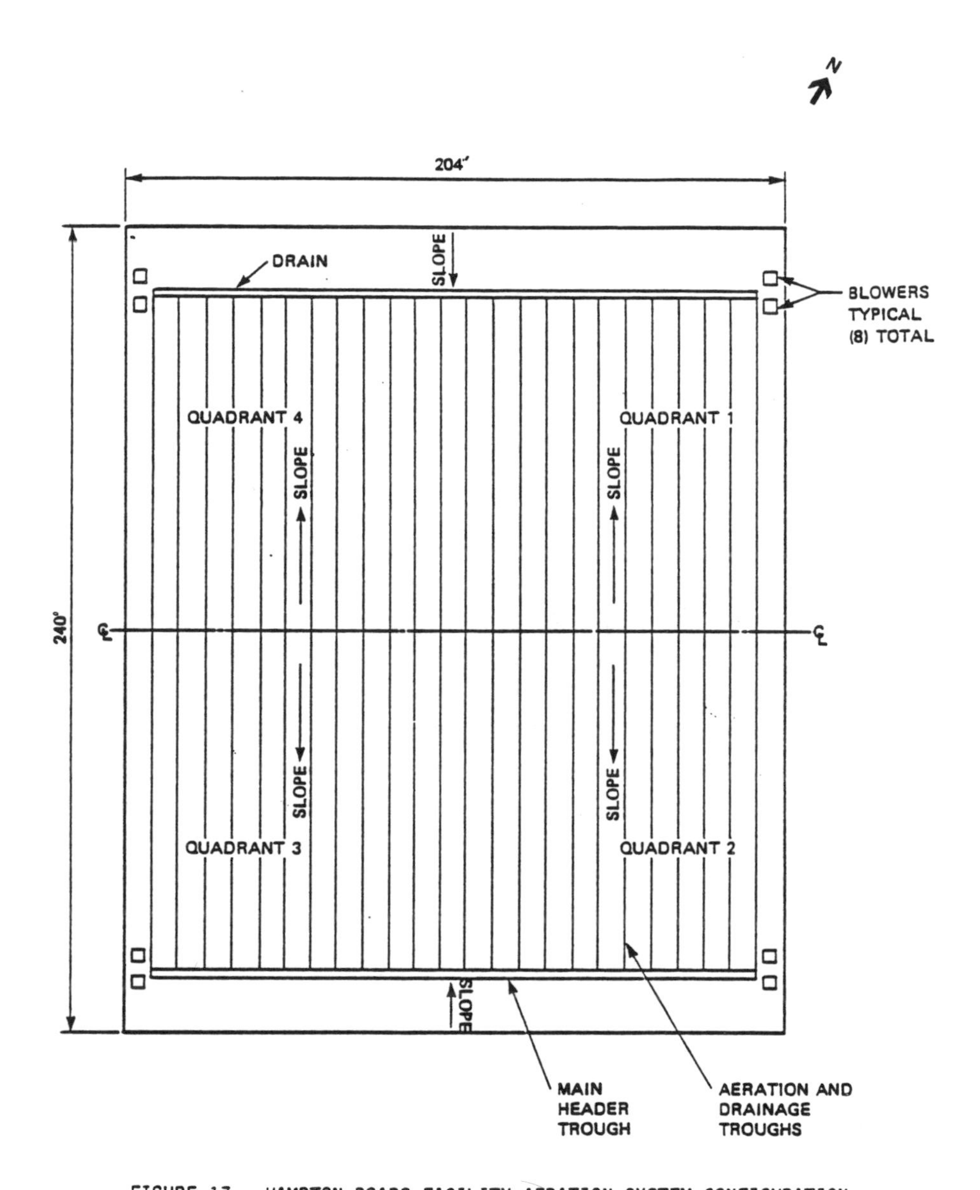

FIGURE 17. HAMPTON ROADS FACILITY AERATION SYSTEM CONFIGURATION

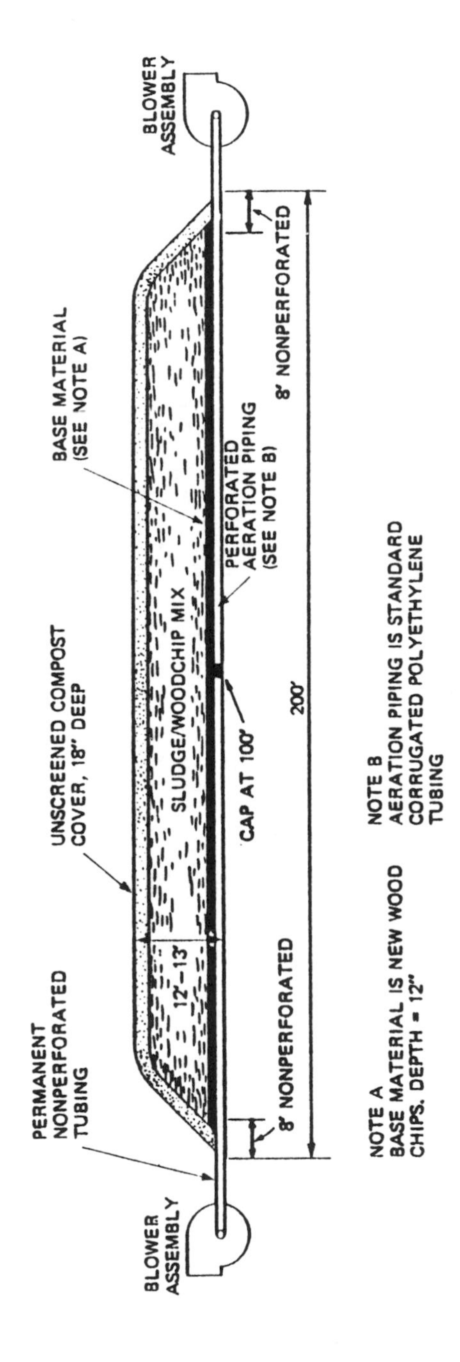

FIGURE 18. PILE CONSTRUCTION AT COLUMBUS FACILITY

chip mixture on the base material in the extended pile configuration; and (4) covering the pile with an 18-inch blanket of screened (or unscreened) compost. The first 12 ft of base material at each end of the extended pile compartment is unscreened compost, and the remaining base material is wood chips. The unscreened compost prevents short-circuiting in the toes, and the wood chips provide uniform aeration within the pile. Two 10-cu yd front-end loaders perform all material handling except for transfer of the mix from the mixing area.

Each extended pile compartment is serviced by one 15-horsepower (hp) blower located at one end of the compartment. Total blower capacity for active composting is 630 hp, which is equivalent to 1.58 hp per wtpd based on a design loading of 400 wtpd. Each blower has a percentage timer and two air pressure meters, one each at the intake and exhaust.

An aeration manifold consisting of four, 6-inch-diameter, flexible plastic hose laterals placed at 5-ft intervals on the paved compost pad extends the length of each compartment. Each aeration lateral consists of 15 ft of solid pipe at the blower end and about 86 ft of perforated pipe. No piping is provided the last 9 ft under the toe farthest from the blowers. For uniform aeration, perforations are tapered as presented below:

Distance from blower, ft	Perforations, square inch per linear foot
0-15	-0-
15-32	0.415
32-52	0.83
52-72	2.66
72-101	4.28
101-110	No pipe

At the Hampton Roads facility (Figures 16 and 17), a 240-ft by 204-ft concrete composting pad is sectioned into quadrants. Each quadrant contains 12 troughs serviced by two 3-hp, variable-speed blowers at each end. Total blower capacity for active composting is 24 hp, which is equivalent to 0.48 hp per wtpd based on a design loading of 50 wtpd. Each trough contains 100 ft of 6-inch, Schedule 40, polyvinyl chloride (PVC) aeration pipe connected to a manifold; one blower services six troughs. The pipe is perforated, with 3/8-inch holes spaced 1 1/2 inches apart, beginning 22 ft from the end of the compost pile. The end of the pipe is capped. The perforation configuration was determined from extensive testing during normal operation and provides the most uniform oxygen levels and temperatures throughout each pile. Airflow is controlled by a damper located at each trough.

Pile construction at this facility includes a wood chip base, 12 to 18 inches deep, except at the toes, and cover material of dry, recycled (unscreened) compost. Cover material depth is 18 inches in the summer and 24 inches in the winter. Base material at the toes is dry, recycled compost to prevent short-circuiting, and 4 to 6 inches of screened compost is placed on the face of the toes for added insulation and to prevent short-circuiting. Each pile is constructed to an initial depth of about 13 ft and widths of 8 to 10 ft.

Preparation of the composting pad for pile construction includes manual cleaning of the aeration troughs; placement of new aeration pipe, if needed; and filling the troughs with wood chips. Pipe replacement, necessitating trough cleaning and placement of new pipe and wood chip filler, has been virtually eliminated since substitution of perforated PVC pipe for flexible, plastic pipe designated in the original design. Whereas the original plastic pipe had to be replaced about every three cycles because crushing caused dead spots in the airflow, PVC pipe now being used typically lasts 2 years or longer.

Time-and-motion studies conducted during the on-site investigation at the Hampton Roads facility indicated the following breakdown of pile construction activities:

	Activity[a]	Time, minutes[b]
1.	7-cu yd front-end loader prepares base material	30
2.	7-cu yd front-end loader transfers compost mix to pile, assisted by 5-cu yd front-end loader	45 (per sludge load)
3.	7-cu yd front-end loader places 18-inch cover material	55
4.	Manual cleaning of troughs and raking of cover material	50

[a]A 5-cu yd front-end loader is sometimes used for these operations.
[b]Per day except as noted.

Pile construction at the Columbus facility (Figure 18) is similar to that at the Site II facility. Aeration piping is laid directly on a concrete compost pad and covered with 12 inches of new wood chips. The perforated piping used is standard 5-inch-diameter corrugated polyethylene tubing which runs the entire length of the pile compartment except for the toe farthest from the blower end. Blowers are 1 hp and are connected to a permanent manifold with two aeration pipes per manifold on 12-ft centers. Total blower capacity for active composting is 40 hp, which is equivalent to 0.20 hp per wtpd based on a design loading of 200 wtpd. A typical extended pile compartment is 200 ft long by 8 ft wide by 12 to 13 ft high. This includes cover material which is typically 18 inches of unscreened, cured compost.

Pile construction at the Columbus facility is performed by several different front-end loader operators. They also perform other activities on site, so it was difficult to develop meaningful time and motion analyses for pile construction activities. Several observations, however, indicated that it took between 24 and 42 minutes, with 30 minutes being a reasonable average, to form a pile containing a mix of 25 tons of sludge and bulking agent.

The assessment of pile construction techniques employed at the three static pile facilities indicates that techniques are quite varied. Pile height for all three facilities is 12 to 13 ft; however, daily compartment lengths of 110 to 200 ft and widths of 8 to 20 ft are used. Base material is typically new wood chips, 12 inches deep, in the central portion of each pile compartment, although one facility reported that depth may vary up to 18 inches. Unscreened compost is used as toe base material at two facilities, whereas new wood chips are used as toe base material at the third facility.

Unscreened compost is used as cover material at all three facilities, and at one, screened compost is also sometimes employed. At two facilities, 18 inches of cover material are used year round, while at the third, 18 inches are employed during the summer months and 24 inches are utilized during the winter months. This latter facility also uses 4 to 6 inches of screened compost on top of unscreened compost cover on the face of the pile toe.

Flexible polyethylene or PVC aeration piping, 5 to 6 inches in diameter is laid directly on paved composting pads at two of the static pile facilities investigated. Six-inch-diameter, rigid PVC aeration piping is placed in troughs cast into the composting pad at the third facility. As noted in Table 10, the number of aeration pipe laterals per daily compartment varies from one to four, the length of unperforated lateral varies from 15 to 22 ft, and the length of perforated lateral varies from 86 to 184 ft. Perforation spacing is uniform at the Hampton Roads facility, which treats anaerobically digested sludge, and 0.88 square inch of perforation area per linear foot of lateral is provided. At the Columbus facility, which treats unlimed, raw sludge, perforation spacing is uniform and 4.00 square inches of area per linear foot of lateral are provided. Variable perforation spacing of 0.415 to 4.28 square inch per linear foot of lateral is employed at the Site II facility, which treats limed, raw sludge.

Unit blower capacity provided for active composting (Table 10) is much greater at the Site II facility (1.58 hp per wtpd) than at the Columbus facility (0.20 hp per wtpd), based on design loadings of 400 and 200 wtpd, respectively, even though both compost raw sludge. Unit blower capacity provided at the Hampton Roads facility, which composts anaerobically digested sludge, is 0.48 hp per wtpd.

Although pile construction techniques are varied, all techniques appear to provide for effective composting based on the on-site observations made during the technology evaluation. Whereas personnel at two of the facilities have made substantial modifications to pile construction techniques relative to those planned at design, the original design configuration is employed at the third (the Columbus facility).

Active Static Pile Composting

The active static pile composting phase, as used in this report, consists of 21 days of aeration for degradation of putrescible sludge organics and inactivation of pathogens. Control of temperature and oxygen content within the piles undergoing active composting is the key operating element required for effective sludge treatment during the composting period. After the active composting period, the pile is torn down and transferred for further processing. Aeration and process control features at each of the static pile facilities and ambient air temperature effects during active static pile composting are described below.

Aeration and Process Control Features--

Key aeration and process control features at the three static pile facilities are summarized in Table 11. The Site II facility, which processes raw, limed sludge, applies the following typical aeration process control parameters for a 21-day active composting period:

1. First week--negative aeration at 3,500 to 4,000 cubic feet per hour per dry ton of sludge (cfh/dt) to promote material decomposition and for odor control.

2. Second and third weeks--aeration is controlled to maintain temperatures between 50 and 60 degrees C, including a minimum of 3 days above 55 degrees C to meet pathogen inactivation requirements; then, aeration is increased to 6,000 cfh/dt for drying and odor control at teardown.

Temperature controllers are installed on each blower, and blower operation is established according to pile temperature. A temperature control set point of 58 degrees C is used. Initially, the aeration blower is off until pile temperature reaches 40 degrees C; then a timer controls blower operation until the temperature set point is reached. During this intermediate period, blower operation is cyclical--5 minutes on and 5 minutes off. After the temperature set point is reached, the blower is on continuously.

Each extended pile compartment at the Site II facility is monitored at five locations for temperature and three locations for oxygen. Thermocouples are left in place during the 21 days of active composting and temperatures are recorded daily. Oxygen readings are taken daily from oxygen-monitoring tubes. A temperature and oxygen monitoring report is maintained for each pile denoting time of measurement, location, and blower mode. Review of routine monitoring reports for eight piles constructed in January, February, July, September, October, and November 1984 showed that the time required to meet the 55 degree C performance criterion varied from 4 to 18 days and that oxygen levels are typically between about 3 and 15 percent.

Piles not meeting the 3-day, 55 degree C performance criterion are recycled into new piles. During August 1984, 2 out of 52 piles did not meet the performance requirements, necessitating reprocessing. The 55 degree C performance criterion is considered to have been met if this temperature is achieved on 3 consecutive days at each monitoring point. It is not necessary that all points meet the criterion at the same time.

During the early stages of active composting of anaerobically digested sludge at the Hampton Roads facility, airflow rate is set to maintain an oxygen content of at least 5 percent and temperatures greater than 55 degrees C within the piles. Typical seasonal operating modes are presented below:

TABLE 11. KEY STATIC PILE AERATION AND PROCESS CONTROL FEATURES

Item	Hampton Roads facility	Site II facility	Columbus facility
Aeration features[a]			
Pre-55 degrees C operation			
Mode	Positive (summer) Negative (winter)	Negative[b]	Positive[b]
Rate, cfh/dt	Summer: 1,100 to 1,200 Winter: 1,000 to 1,100	3,500 to 4,000	-[c]
Blower cycle	10 minutes on, 20 minutes off	5 minutes on, 5 minutes off[d]	20 minutes on, 10 minutes off[e]
Oxygen content, percent	≥ 5	≥ 5	Not used for control
Post-55 degree C operation[b]			
Mode	Positive	Negative	Positive
Rate, cfh/dt	2,500 to 2,800	6,000	-[f]
Blower cycle	Continuous	Continuous	Continuous
Oxygen content	Approaches maximum	Approaches maximum	-
Blower control procedure	Manual	Automatic	Manual
Temperature monitoring			
Technique	Probes left in pile; read daily	Probes left in pile; con- nected to blower controls	Periodic temperature probe readings
Number of points	6	5	Random
Oxygen monitoring			
Technique	Monitoring tubes, daily readings	Monitoring tubes, daily readings	Not monitored
Number of points	3	3	-

[a]Pre-55 degree C operation applies prior to piles reaching 55 degrees C for 3 consecutive days.
Post-55 degree C operation applies after piles reach 55 degrees C for 3 consecutive days.
[b]Operation does not change with season.
[c]Rate is determined by operators.
[d]Blower cycle is not started until 40 degrees C is reached.
[e]During the summer months blower cycle starts after pile construction. During the winter months blower cycle starts after 40 degrees C is reached.
[f]Blower rate is set at 100 percent.

1. Winter operation--negative aeration, with blowers cycled 10 minutes on and 20 minutes off at 1,000 to 1,100 cfh/dt.

2. Summer operation--positive aeration, with blowers cycled 10 minutes on and 20 minutes off at 1,100 to 1,200 cfh/dt.

For both operating modes, once the 55 degree C temperature criterion is met for 3 consecutive days, the blowers are put on positive aeration continuously at 2,500 to 2,800 cfh/dt.

In the negative aeration mode, exhausted air must be treated for odors by filtering through a finished compost pile. The positive aeration mode requires no odor control; additionally, in this latter mode, drying of compost to 55 percent total solids is sometimes achieved during dry weather, enabling direct screening of the compost after teardown. With the positive aeration mode, leachate does not form as it does with negative aeration. Because of this, material in the aeration troughs does not get wet under positive aeration, and odor is not generated. Periodic trough cleaning is still required with either aeration mode, however.

Each daily pile at the Hampton Roads facility is monitored at six locations for temperature and at three locations for oxygen. Thermistors are left in place during the 21 days of composting. Temperatures are recorded daily until 55 degrees C are obtained for 3 consecutive days, and then once per week. Oxygen readings are taken daily using a portable oxygen meter.

Temperature and oxygen monitoring data for two piles constructed with wood chip/sludge mixtures prepared using different mixing techniques were analyzed as part of the technology evaluation. One pile constructed June 26, 1984, was mixed using a front-end loader and rototiller. A second pile constructed June 27, 1984, was mixed only by a front-end loader. The standard summertime blower cycle was maintained for the pile constructed June 27 (front-end loader mix). However, for the pile constructed June 26 (front-end loader and rototiller mix), blowers were cycled 15 minutes on and 15 minutes off for the first 2 days of composting, before returning to the standard cycle. Positive aeration was used for both piles throughout the 21-day active composting period.

In both piles tested, temperatures greater than 55 degrees C were achieved within 3 to 4 days and stayed well above 55 degrees C (up to 90 degrees C) until application of high-rate, positive aeration caused a temperature drop. Temperatures were relatively uniform throughout the piles, varying about 5 to 10 degrees at any location. Typically, oxygen levels increased from 0 to 10 percent during the first 6 days of aeration, decreased to 6 percent in the next 6 days of operation, and reached atmospheric (20 percent) once high-rate positive aeration commenced. Comparison of data for the pile constructed with compost mixed by front-end loader and a rototiller to that for compost mixed only by front-end loader did not show any differences in temperature and oxygen characteristics.

The Columbus facility, which processes unlimed, raw sludge, uses blowers equipped to control the composting operation either by a timer, temperature probe in the compost pile, or manually. However, blowers are not currently attached to temperature probes because of mechanical difficulties. Thus, manual control, in conjunction with a timer, is utilized for process control.

Process control varies seasonally. In winter, aeration is not started until the pile reaches 40 degrees C; then it is on for 20 minutes and off for 10. In summer, aeration starts as the pile is constructed. Positive aeration is used. The process is monitored by the use of temperature probes that are inserted into the pile as the pile is set up. A reading is taken and recorded and this monitoring is continued until three daily readings of 55 degrees C or greater are recorded. At such time, blower rate is increased to 100 percent to remove moisture from the material. Oxygen content is not routinely measured for process control at the Columbus facility.

Ambient Air Temperature Effects--

Temperature monitoring data from the Hampton Roads, Columbus, and Site II facilities were reviewed as part of the technology evaluation to determine whether the time required to reach 55 degrees C could be correlated with ambient temperature. For piles constructed January through March 1983 and June through August 1983 at the Hampton Roads facility, the analysis showed that, on the average, 7 to 9 days were required to reach 55 degrees C when ambient temperatures were between 6 and 19 degrees C (January-March). On the average, 3 to 5 days were required when ambient temperatures were between 24 and 29 degrees C (June-August). Of 119 piles tested, 49 piles required from 5 to 7 days to reach 55 degrees C for 3 consecutive days. Twenty-seven piles required 3 to 5 days to meet the standard, and 28 piles required 7 to 9 days to meet the standard. A more complete discussion of the analysis is given in Appendix B.

Analysis of data from the Columbus and Site II facilities was less rigorous than that from the Hampton Roads facility. For piles constructed during the months of January through April 1984 at the Columbus facility, a definitive time-temperature relationship could not be identified. However, the following general responses were observed:

1. During the first 16 days in January, ambient temperature was decreasing from 2 degrees C to values in the vicinity of -26 degrees C or less. About 7 to 9 days were required for piles constructed during this period to reach 55 degrees C.

2. For an 11-day period in the latter part of March, ambient temperatures were in the range of about 1 to 7 degrees C. The number of days to reach 55 degrees C decreased steadily from 9 to 4 during this period.

3. By mid-April, when ambient temperatures exceeded 4 degrees C, 1 to 3 days were required to reach 55 degrees C.

4. The maximum number of days required to reach 55 degrees C was never greater than 14 days during the 4-month period.

Review of routine temperature monitoring data from eight Site II piles constructed in January, February, July, September, October, and November 1984 showed that although a minimum time of 4 to 7 days was required to reach 55 degrees C, during January, February, and October, longer times were sometimes required (up to 18 days in January).

The results of the temperature response analysis at the three static pile facilities demonstrate the influence of ambient temperature on the time required to reach 55 degrees C during active static pile composting. Although generalizations are difficult because of different aeration rates, process control procedures, sludge types, and pile construction techniques employed at these facilities, the comparison does show that (1) ambient temperature influences the time required to reach 55 degrees C, and (2) the 55 degree C performance criterion can be met within 21 days even under widely varying composting conditions.

Postcomposting Operations

Postcomposting operations include pile teardown, drying, screening, and curing steps. An assessment of these operations and finished compost quality and distribution at the three static pile facilities investigated is presented in this section.

Pile Teardown--

After the 21-day active static pile composting period, each compost pile is torn down, typically by front-end loader, and composted material is transported to a screening, drying, or curing area. At the Hampton Roads facility, pile breakdown is not performed during wet weather or in early morning hours when air inversions may result in odor generation. At the Site II facility, piles are aerated for 24 to 48 hours prior to teardown in order to reduce temperatures and minimize steam or odor release. At the Columbus facility, piles are supposed to be torn down after a 21-day active composting period. However, during the on-site investigations conducted as part of the technology evaluation, piles were left over 30 days before being torn down; according to facility personnel, this is a function of space on the pad and equipment availability. Piles that were torn down were quite wet and had some odor. However, the odor dissipated a short distance from the pile.

The assessment of pile teardown techniques employed at each static pile facility demonstrates that this operation can be a source of odors. To minimize the potential impact of this operation, piles can be aerated at high rates prior to pile teardown. Alternatively, materials management operations can incorporate sufficient process area at the site to provide for periods when weather conditions are not conducive to teardown activities.

Drying and Curing--

Material composted by the aerated static pile method often is not dry enough to efficiently screen directly. In these cases, a separate drying step is employed. However, as previously noted, personnel at the Hampton Roads facility have found that during the summertime, when positive aeration occurs for the entire 21-day period, composted material may be dry enough for screening, and the drying step is bypassed. Screened compost is then cured to provide additional pathogen inactivation and odor removal prior to storage for distribution. Benefits of this mode of operation are immediate recycling of wood chips and reduction in the area required for curing. At other times, however, separate drying is performed at this facility and, in these cases, unscreened compost is then stored in a curing area until weather permits drying. Drying operations entail transfer of cured, unscreened compost by front-end loader to one of two drying slabs (see Chapter 3) where the material is spread out to a depth of 15 to 18 inches and periodically rototilled until a minimum of 50 percent total solids is achieved. Total slab area available for drying is about 1.7 ac, or 1,500 sq ft per wtpd. Typically, drying time is less than 3 days in the summertime when temperatures are high. After drying, material is piled under a covered shed until screened.

As designed, drying at the Hampton Roads facility was to be performed in a drying shed, now the mixing and screening shed. Drying was to be accomplished by negative aeration of 3-ft-tall windrows; however, this method was not effective. Plant personnel tried drying by spreading compost out to a 3-inch depth with a manure spreader, but this method was time-and space-consuming. Facility personnel report that the current method is the most cost-effective as it requires the least materials handling.

Overall curing time at the Hampton Roads facility is a minimum of 30 days, as required by state regulations. Additional storage time is used, however, in both the curing area and in the finished compost storage area, depending on materials handling considerations. An area of 1.6 ac is available for curing/storage of unscreened compost. This area is equivalent to about 1,400 sq ft per wtpd.

At the Site II facility, each pile is torn down by a front-end loader after the active compost period, and composted material is transported to a drying area where it is restacked for an aerated drying step prior to screening (prescreen drying). The area for prescreen drying has been provided by adapting one-half of a covered active composting pad as described in Chapter 3. The prescreen drying area occupies 0.8 ac, which is equivalent to about 100 sq ft per wtpd. Compost is restacked over 4-inch perforated pipe and connected to 1-hp blowers. Constant aeration is applied. The same pile dimensions as those for active composting are used for prescreen drying operations.

Temperature and oxygen are monitored during prescreen drying, and generally pile temperatures increase rapidly to 55 degrees C after restacking (within 1 day) before dropping to ambient. Oxygen content is maintained at >15 percent during the prescreen drying step. Moisture reduction data from 18 piles undergoing prescreen drying during July and August 1985 are summarized below:

Drying period, days	Number of piles	Average moisture, percent		
		Initial	Final	Reduction
1-3	3	46.0	43.6	2.4
4-6	7	46.2	42.4	3.8
7-12	4	47.5	42.1	5.4
13-17	4	47.7	41.5	6.2

The tabulated data show that under the conditions studied at the Site II facility, moisture content was reduced by about 3 to 6 percent depending on the drying period employed.

After screening, front-end loaders transport screened compost to storage for a minimum 30-day aerated curing period. Two uncovered concrete pads, each 120 ft by 600 ft, are provided for storing screened or unscreened compost. A roadway, 50 ft wide, is provided around and between these pads. The aerated curing operation occupies an area of 3.3 ac, which is equivalent to about 400 sq ft per wtpd, excluding area for the access roads. The aerated curing pads are sloped toward the roadways where catch basins are located for stormwater collection and transfer to a containment pond. Maximum storage capacity of this area is limited to a 6-month production of screened compost.

One-hp blowers are used for curing pile aeration and aeration pipes are 4-inch laterals, three per blower, located on 8-ft centers. Aeration maintains high oxygen levels in the curing piles and prevents release of odors. Aeration continues until the compost is distributed. Temperature and oxygen are monitored during the aerated curing period, and, as with prescreen drying, pile temperatures generally increase rapidly to 55 degrees C within 1 day after restacking before dropping to ambient. Oxygen content is maintained at >15 percent during aerated curing.

As part of recent facility modifications at the Columbus facility, conveyors are being installed to facilitate movement of materials on the site. However, front-end loaders are now being used. Compost needs to be cured prior to screening because of moisture and odor problems, and an uncovered, aerated curing pad, approximately 4.6 ac in size, is provided for this purpose. The aerated curing area at the Columbus facility is equivalent to about 1,000 sq ft per wtpd. Aeration provided to the curing piles is essentially the same as used in the active composting phase. Blowers are on a positive mode 100 percent of the time and 5-inch perforated pipe is used for the aeration step.

After the material is cured over aeration for at least 30 days, it is used in one of four ways. It can be recycled as part of the bulking agent, it can be used as insulation, it can be further dried by restacking, or, if dry enough, it can be screened. During the on-site investigations which were conducted during the winter months (December-February), only the first and second options were observed. A front-end loader was used to break down all curing piles.

Screening Operations--

Screening is an important step in the static pile process as this operation is necessary for effective recovery and recycling of bulking agent (wood chips). Finished compost uses typically require screens which produce finished compost grades in the 1/4- to 3/8-inch size range, and, under optimum conditions, wood chip recoveries of 80 to 90 percent[7] can be obtained in

[7]Wood chip recoveries are based on unscreened compost volume.

these size fractions using the types of screening equipment employed at the three aerated static pile facilities investigated. To do so requires that the total solids content of unscreened compost be between 50 and 60 percent, depending in part on screen size and design. Exceeding the upper limit can result in losses from excessive dust generation, while not meeting the lower limit inhibits the separation of fines from the wood chips. The latter situation leads to lower wood chip recoveries, reduced screening rates, and often mechanical problems with the screening equipment.

Under sustained routine operation, wood chip recoveries of 65 to 85 percent are typical given the heterogeneity of compost material, operational variability, and other factors. As shown in Table 12, three replicate observations made during the technology evaluation at the Hampton Roads facility using a 1/4-inch screen at screening rates of 99 to 107 cubic yards per hour (cu yd/hr) yielded bulking agent recoveries of 64, 76, and 86 percent (average = 75 percent). Initial total solids of the unscreened compost was 56 percent. By comparison, review of facility records for 3 months of routine operation showed recoveries of 42 to 93 percent, with an average of 64 percent. Unscreened compost total solids content for these months was between 46 and 79 percent.

Two mobile power screens equipped with 13-cu yd hoppers produce, on the average, about 550 cu yd of screened compost per month at the Hampton Roads facility. Screening rates vary between 60 cu yd/hr and 120 cu yd/hr during routine operations, depending on the dryness of the compost. Screening performance tests conducted by facility personnel show that wood chip recovery is best (80 to 90 percent) when compost contains 55 to 60 percent total solids, although other factors affect performance as noted above. The screens can be operated at a total solids content as low as 50 percent; however, the screen deck clogs rapidly and requires constant cleaning.

At the Site II facility, which employs fully enclosed screening equipment, wood chip recoveries of 85 to 87 percent can be consistently achieved during dry weather, as shown in Table 13. However, under wet-, cold-weather conditions, screening problems occur and recoveries are less even with effective moisture control in the incoming unscreened compost.

Screening at the Site II facility is accomplished in a 210-ft by 200-ft fully enclosed building. Unscreened compost is discharged into 30-cu yd hoppers which feed the material onto conveyor belts that lead to three vibratory screens. Each screen is designed to process up to 120 cu yd of material per hour. On top of the three screens are vacuum hoods which are intended to collect any dust generated or odor released and carry it out of the building through pipes to a compost filter. Chips that remain on top of the screens are conveyed out one side of the building for storage and recycle. Screened compost falls onto another conveyor belt system underneath the screens and is carried out another side of the building where it is piled under cover prior to transfer for curing.

Screening performance varies depending on the moisture content of unscreened compost, equipment malfunctions, and season. If unscreened compost contains more than 50 percent moisture, the screens clog rapidly reducing

TABLE 12. SCREENING PERFORMANCE AT THE HAMPTON ROADS FACILITY

Unscreened compost							
Total solids, percent	Volume cu yd[a]	Screening rate, cu yd per hour[b]	Elapsed time, minutes	Recycled wood chip volume, cu yd[c]	Screened compost volume, cu yd[c]	Volume increase, percent[d]	Wood chip recovery, percent[e]
56	25	107	14	19	9	12	76
56	14	105	8	12	7	38	86
56	28	99	17	18	14	14	64

[a]Measured by counting the number of front-end loader bucket loads deposited into screen hopper.

[b]Screen size--1/4 inch.

[c]Calculated from volume of cone of material deposited after screening.

[d]Change in total volume (recycled wood chips plus screened compost) due to density change through screening.

[e]As percent of unscreened compost volume.

TABLE 13. DRY WEATHER SCREEN PERFORMANCE AT THE SITE II FACILITY

		Average volume, cu yd/day		
Month, 1984	Unscreened compost	Recycled wood chips	Screened compost	Wood chip recovery, percent
July	1,300	1,100	200	85
August	1,400	1,220	180	87
September	1,413	1,233	180	87
October	1,320	1,127	193	85

Note: Data from operating records.

screening rate and wood chip recoveries. Downtime for the equipment is also considerable. During August and September 1984, two screens were out of service, one typically 5 to 6 hours per day and another 1 to 2 hours per day. In October 1984, one screen was out of service for 11 days, two others were out of service 1 to 2 days.

At the Columbus facility, compost is screened either in a mobile, rotary screen with a 1/4-inch mesh or a fixed, vibratory screen with a 3/8-inch mesh. The screens are specified for 150 and 240 cu yd/hr, respectively, at 45 percent moisture. The 1/4-inch screened material is recommended for use as top dressing for existing lawns, golf course tees and fairways, and for other uses in which the appearance of wood chips is undesirable. The 3/8-inch material is recommended for use in building lawns, general landscaping, mixing into potting soil for horticultural uses, and as a topsoil replacement.

Moisture has been an ongoing operational problem at the Columbus facility. Screening was not being performed during the on-site investigations, and facility personnel report that this is typical. As noted in Chapter 3, a large area (10 ac) is provided at this site for unscreened compost storage during winter months. Although the screens were not in operation during the technology evaluation, a manual screening test to assess potential performance was conducted on two random samples of unscreened compost having an average total solids content of about 59 percent. Based on average values from the tests, the following results were obtained:

	Quantity of screened material, percent	
Size of screened material	By weight	By volume
>3/8 inch	80	80
1/4 to 3/8 inch	10.6	11.6
<1/4 inch	9.4	8.2

These limited data indicate that screen performance at the Columbus facility, if operationally feasible on a routine basis, could yield wood chip recoveries of 80 to 90 percent.

The screening operations assessment at the Hampton Roads, Site II, and Columbus facilities demonstrates the need to define screening effectiveness, including expected variability, for aerated static pile applications. Although optimum screen performance (80 to 90 percent wood chip recoveries) can be approached in some cases, sustained performance during routine operations is typically less based on the experience at the three static pile facilities investigated. Key factors affecting performance include weather, equipment malfunctions, and drying capability prior to screening. As previously noted, screening recoveries of 65 to 85 percent based on unscreened compost volume appear to be achievable, on average, for routine plant operations.

Finished Compost Quality--

Table 14 presents representative finished compost characteristics at the three aerated static pile facilities investigated during the technology evaluation. Total solids reported are between 51 and 69 percent, with the higher value being reported by the Columbus facility. Finished compost is stored under cover at this location, whereas at the other two sites, open storage is used. Total volatile solids of finished compost produced from raw sludge (without lime) is slightly higher than that produced from digested sludge; finished compost bulk densities are between 900 and 1,500 lb/cu yd based on the data presented.

Monitoring of finished compost at the Hampton Roads facility includes monthly measurements for heavy metals and nutrients and semiannual analysis of minerals and metals. Additionally, coliform and Salmonella testing is performed monthly. Salmonella has never been detected in the finished compost, and coliform levels are within acceptable limits (less than 60 colonies per 100 grams). Independent microbiological analyses of finished compost (duplicate samples) during the on-site investigation yielded no Salmonella and <2 fecal coliform colonies per 100 grams, confirming these data. Metals content of finished compost has consistently been within the required range for land application. Independent analysis of finished compost during the technology evaluation indicated values higher than the range experienced previously by the facility but still within acceptable limits.

Finished compost at the Site II facility is sampled every 2,000 cu yd for seven heavy metals, nitrogen, phosphorus, and potassium. Also, Salmonella levels are measured to monitor for pathogen inactivation. Due to low heavy metals content of Blue Plains sludge, these constituents are not a concern for compost distribution. Heavy metals have always been well under allowable limits. Salmonella has never been detected in the 21-day unscreened or finished compost. Additionally, except for low levels of potassium, nitrogen and phosphorus, the compost makes an excellent soil conditioner. Compost for greenhouse use sometimes requires that soluble salts present from sludge chemical conditioning be leached out prior to use.

Finished compost at the Columbus facility is sampled monthly by plant personnel and analyzed for plant nutrients, heavy metals, and Salmonella, as well as total and total volatile solids. As shown in Table 14, Salmonella has not been detected at this facility. A comparison, based on plant records, of finished compost quality for 1/4-inch screened compost and that for 3/8-inch compost, indicates that variations in metals and nutrient content between the two sizes are not significant, as shown in Table 15.

Finished Compost Distribution--

Finished compost is sold to local users at all of the facilities investigated. Demand varies with season and thus materials management procedures for low-use periods are employed. The price of finished compost varies from about $3 per cu yd to $9 per cu yd, depending on facility location and quantity purchased. Typical finished compost production rates are about 0.5 to 1.0 cu yd per wet ton of sludge processed, based on performance at the

TABLE 14. FINISHED COMPOST CHARACTERISTICS FOR STATIC PILE FACILITIES

Constituent[a]	Concentration range		
	Hampton Roads facility[b]	Site II facility[c]	Columbus facility[d]
Solids, percent			
Total	52[e]	51	61-69
Total volatile	49[e]	-	46-50
Bulk density, lb/cu yd	900[e]	1,000-1,500	1,100-1,400[e]
Trace minerals, mg/kg			
Cadmium	5-10	2.0	7-30
Chromium	68-123	-	180-350
Copper	273-448	112	200-300
Lead	75-225	40	190-340
Mercury	-	0.4	-
Nickel	23-43	46	55-170
Zinc	578-1,039	176	1,400-2,200
Nitrogen, as N, percent			
Total Kjeldahl	-	0.9	1.0-1.5[f]
Organic	1.4-2.1	-	-
Ammonia	0.5-0.9	0.1	-
Total phosphorus, as P, percent	1.8-2.8	0.6	2.8-4.1
General minerals, mg/kg			
Calcium	7,875-21,250	-	-
Iron	16,375-26,000	-	-
Magnesium	2,875-4,000	-	-
Potassium	588-3,500	-	-
Soluble salts, as $CaCO_3$, percent	-	7.6	-
Microbiological, number/100 gm			
Fecal coliform	<60	-	>100[e]
Salmonella	Not detected	Not detected	Not detected
pH	5.9-6.8	-	4.6-9.0[e]

[a]Dry weight basis, except density and pH.
[b]As reported by the facility for 1983, except as noted.
[c]As reported by the facility for July 10, 1984.
[d]As reported by the facility, except as noted.
[e]Independent analyses during technology evaluation.
[f]Total nitrogen.

TABLE 15. EFFECT OF SCREEN SIZE ON FINISHED COMPOST QUALITY AT COLUMBUS FACILITY

Constituent[b]	Facility data[a]	
	1/4-inch screen	3/8-inch screen
Trace minerals, mg/kg		
Cadmium	17	18
Chromium	300	289
Copper	269	247
Lead	251	241
Nickel	69	73
Zinc	1,818	1,704
Total nitrogen, as N, percent	1.4	1.2
Total phosphorus, as P, percent	3.3	3.3

[a]Average of values reported.
[b]Dry weight basis.

Hampton Roads and Site II facilities. The Columbus facility does not screen for long periods; therefore, meaningful sustained production figures are not presented.

To introduce the compost product to potential customers, compost from the Hampton Roads facility was initially given away to the general public. However, since 1983, the facility has sold the compost on a bulk basis at $7.50 per cu yd for quantities less than 10 cu yd and $6 per cu yd for quantities greater than 10 cu yd. Additional discounts apply to purchases above 1,000 cu yd. Sales revenue for 1983 showed that general landscapers accounted for 38 percent of sales; the general public, 37 percent; and the remaining 25 percent from nurseries, institutions, and municipalities. The facility has an aggressive marketing program administered by a part-time marketing agent. The agent also provides technical expertise to customers and works to educate the general public on the beneficial uses of the compost. Sales for the first 8 months of 1984 were twice those for the same period in 1983. An attractive and informative brochure sent to potential customers (and distributed with the compost) has been an effective form of advertising. In 1984, a private company bagged the compost for retail sale.

Finished compost from the Site II facility is marketed by Maryland Environmental Services under the trade name ComPRO for use as a soil amendment and conditioner. Currently, the product is sold in bulk, with delivery service provided (no individual distribution on site). Additionally, ComPRO is now also distributed in small bagged quantities through more than 75 retail establishments from nurseries to hardware stores. The fall 1984 bulk price schedule was as follows:

Quantity, cu yd	Cost, dollar/cu yd
0-100	4.00
101-200	3.80
201-400	3.60
401-600	3.40
>601	3.00

A 5-cu yd minimum purchase is required for bulk purchase; transportation charges vary between $3 and $9 per cu yd depending on the distance from the site and haul truck size (20-cu yd and 40-cu yd trailer trucks are available). Finished compost has been approved for unrestricted use (except for tobacco production). In order to apply compost to agricultural croplands, a readily obtainable permit is required.

Sales of ComPRO have totaled 55,000 cu yd between April 1983 (start of operations) and November 1984. In October 1984, 4,920 cu yd were distributed; users included garden centers, landscapers, municipalities, developers and the general public, with quantities per user ranging between 2 and 233 cu yd. Advertising is accomplished by word of mouth and radio commercials. Retailers of ComPRO also advertise in local newspapers and magazines. Research on uses and benefits of ComPRO is carried out at the University of Maryland.

Finished compost from the Columbus facility is marketed as "COM-TIL". Initially, it was distributed at no cost (between September 1, 1981, and September 21, 1983) during a trial program to establish a marketing strategy for ongoing distribution. During 1983, the last year with complete data, 4,987 tons of compost were distributed, 21 percent of which was 1/4-inch compost and 79 percent of which was 3/8-inch compost. A breakdown of compost distribution by user category for 1983 is presented below:

Compost user	Quantity, percent of total
Governmental agencies	38
Golf course industry	30
Compost related businesses	16
Public relations/private	13
Church/cemetary/institutions	4

The basic cost of finished compost has been established at $9 per cu yd or $18 per ton, with reductions for large-volume users on a case-by-case basis. Columbus facility staff are responsible for all compost sales.

General Facility Operations

A discussion of material flow considerations; odor control; aerosol measurement and control; and leachate, condensate, and runoff control is presented in this section.

Material Flow Considerations--

A comparison of material flow information obtained by independent testing at the Hampton Roads and Columbus facilities is presented in Tables 16 and 17. The information provided represents a snapshot of material flows based on samples obtained during the on-site investigations.

The data in Table 16 show that, at a similar total solids content (16 percent), dewatered raw sludge from the Columbus facility is about 25 percent more dense and is 31 percent higher in volatile content than dewatered, anaerobically digested sludge from the Hampton Roads facility. New wood chips at each of the facilities have similar bulk densities (456 and 671 lb/cu yd) and total solids content (56 and 58 percent), but volatile content is higher in the new chips used at the Columbus facility. Possibly, the new wood chips used at the Columbus facility are fresher than those used at the Hampton Roads facility.

Recycled wood chips (after screening) employed at the Hampton Roads facility have a bulk density of about 770 lb/cu yd, a total solids content of 55 percent, and a volatile content of 82 percent. Because of screening problems at the Columbus facility during the on-site technology evaluation investigations, unscreened wood chips were being recycled as a bulking agent. The material employed had a bulk density of 1,087 lb/cu yd, a total solids content of about 44 percent, and a volatile content of 66 percent. Some screened wood chips were available on site, and these were analyzed even

TABLE 16. CHARACTERISTICS OF MATERIALS FOR STATIC PILE MATERIAL FLOWS

Process material	Bulk density, lb/cu yd		Total solids, percent		Total volatile solids, percent	
	Hampton Roads facility	Columbus facility	Hampton Roads facility	Columbus facility	Hampton Roads facility	Columbus facility
Dewatered sludge	1,544	1,943	16.0	15.8	54.0	71.0
Wood chips						
New	456	671	56.0	58.0	71.0	98.0
Recycled[a]	769	1,087	.55.0	44.4	82.0	66.0
Sludge/wood chip mix	830	1,230	41.0	38.8	69.0	72.7
Unscreened compost	933	1,123	48.0	39.8	68.0	62.7
Dried, cured compost	875	1,120	56.0	40.8	59.0	66.0
Screened compost	866	1,164	52.0	51.6	49.0	63.5

[a]See text.

Note: The Hampton Roads facility processes anaerobically digested sludge. The Columbus facility processes a mixture of unlimed, raw primary, and secondary sludges.

TABLE 17. MATERIAL FLOWS FOR STATIC PILE FACILITIES

Process material	Dry weight, tons		Wet weight, tons		Volume, cu yd	
	Hampton Roads facility	Columbus facility	Hampton Roads facility	Columbus facility	Hampton Roads facility	Columbus facility
Dewatered sludge	1.0	1.0	6.3	6.3	8.0	6.5
Wood chips						
New[a]	-	1.0	-	1.8	-	5.3
Recycled[b]	6.1	6.0	10.9	13.5	28.2	24.9
Sludge/wood chip mix	7.0	8.0	17.0	21.6	41.0	35.1
Unscreened compost	6.9	7.3	14.5	18.2	30.9	32.4
Dried, cured compost	5.4	7.1	9.6	17.3	22.0	30.9
Screened compost	1.0	0.2	1.8	0.4	4.3	0.7
Screened wood chips	4.4	0.9	7.8	1.7	20.2	5.0

[a]New wood chips were not being used as a bulking agent at the Hampton Roads facility when the materials flow data were obtained.

[b]See text.

Note: Based on data collected during the technology evaluation. Numbers may not balance because of rounding and inclusion of cover and base materials in some instances.

though their age or origin were not known. The screened wood chips had a bulk density of 677 lb/cu yd, a total solids content of 51.6 percent, and a volatile content of 77.5 percent.

Data presented in Table 16 further show that the sludge/wood chip mix at the Columbus facility is more dense than that at the Hampton Roads facility, as would be expected because of the different sludge densities. Note, however, that the total solids content of the mix at the Columbus facility is below 40 percent, whereas that at the Hampton Roads facility is slightly above 40 percent.

Material flows presented in Table 17 are normalized on a dry-ton-of-sludge basis. As noted, numbers may not balance because of rounding and inclusion of cover and base materials in some instances. The volume per dry ton of dewatered, raw sludge processed at the Columbus facility (6.5 cu yd) is less than that of the dewatered, anaerobically digested sludge processed at the Hampton Roads facility (8.0 cu yd) because of the density difference noted above.

The normalized material flows show that 28.2 cu yd of bulking agent are employed per 6.3 wet tons of dewatered sludge treated at the Hampton Roads facility. This is equivalent to a mix ratio of 4.5:1 (cu yd:wet tons). For the same wet weight (6.3 tons) of sludge treated at the Columbus facility, 30.2 cu yd of bulking agent are employed. This is equivalent to a mix ratio of 4.8:1. On a volume-to-volume basis, corresponding mix ratios are 3.5:1 at the Hampton Roads facility and 4.7:1 at the Columbus facility. Material flow data were obtained under summertime conditions at the Hampton Roads facility and under wintertime conditions at the Columbus facility.

The normalized material flows further show that the Hampton Roads facility employs 10.9 wet tons of bulking agent per dry ton of sludge compared to 15.3 wet tons of bulking agent per dry ton of sludge at the Columbus facility. Thus, the wet weight of bulking agent handled at the Columbus facility is 40 percent greater than that handled at the Hampton Roads facility. On a volumetric basis, however, the difference is less. Bulking agent volumes per dry ton of sludge processed at the Columbus and Hampton Roads facilities are 30.2 cu yd and 28.2 cu yd, respectively, and on this basis, the difference is only 7 percent. The large percentage difference on a wet-weight basis (40 percent) is undoubtedly due to moisture present in the recycled (unscreened) wood chips employed at the Columbus facility.

Based on the material flows presented in Table 17, the Hampton Roads facility produced 1 dry ton of screened compost per dry ton of sludge, while the Columbus facility only produced 0.2 dry ton of screened compost per dry ton. On a volume basis, the Hampton Roads facility produced approximately 0.5 cu yd of screened compost per cu yd of sludge, while Columbus produced 0.1 cu yd of screened compost per cu yd of sludge. Because of screening problems at the Columbus facility, the production value for this facility is not considered to be representative of steady state operation.

Odor Control--

Odor control features are a major aspect of operations at all three static pile facilities investigated. At the Site II facility, key design features for odor control included (1) a centralized compost filter system for cleaning exhaust gas from pile aeration, (2) a completely enclosed screening facility with a separate exhaust gas filter pile, and (3) a leachate and condensate collection and disposal system. Key operational features for odor control include (1) management of initial mix total solids content and uniformity, (2) modified pile construction techniques to ensure uniform aeration, (3) aeration rate control during active composting and prescreen drying to ensure that the material does not go anaerobic, (4) high rate aeration for 24 to 48 hours prior to pile teardown, and (5) effective site housekeeping.

A schematic of the compost filter system used at the Site II facility is shown on Figure 19. This system has not been effective in removing odorous compounds that are transported in the exhaust gas from the active piles for two reasons: (1) Water sprays which are intended to cool the exhaust air and remove soluble compounds by condensation have not worked effectively, and (2) odorous compounds that reach the filter piles are not absorbed by the filter medium. Because of this, a new exhaust gas scrubber system has recently been installed and is undergoing testing to determine its effectiveness.

The effectiveness of odor control features at the Site II facility prior to installation of the new scrubber system was reviewed as part of the technology evaluation. Thirteen complaints were received in August 1984 on 6 separate days, while 40 were received in September on 7 separate days. However, the majority of the complaints received in September (78 percent) were received on 2 days: September 2 (16 complaints) and September 13 (15 complaints). These dates generally coincide with dates when sour piles were torn down and recycled. In the 4-month period of operation between July and October 1984, piles were recycled only on these days. Thus, although improvements in odor control are needed, and planned, it appears that the most severe odor problems have been associated with upsets in the composting process and not with routine operations.

Ammonia, hydrogen sulfide, methyl and ethyl mercaptans, and sulfur dioxide concentrations were measured with a Matheson-Kitagawa gas sampling apparatus during the on-site investigation at the Hampton Roads facility. Ammonia was detected at several locations tested; however, other gases were not. Except for ammonia, no specific odors were identified by participants in the on-site investigation.

Ammonia concentrations, in parts per million (ppm), measured 1 to 3 inches above the surface of various materials in the composting process train were as follows:

Sample location	Ammonia concentration, ppm
Dewatered sludge, as delivered	2.5
Active pile, 14 days old	5

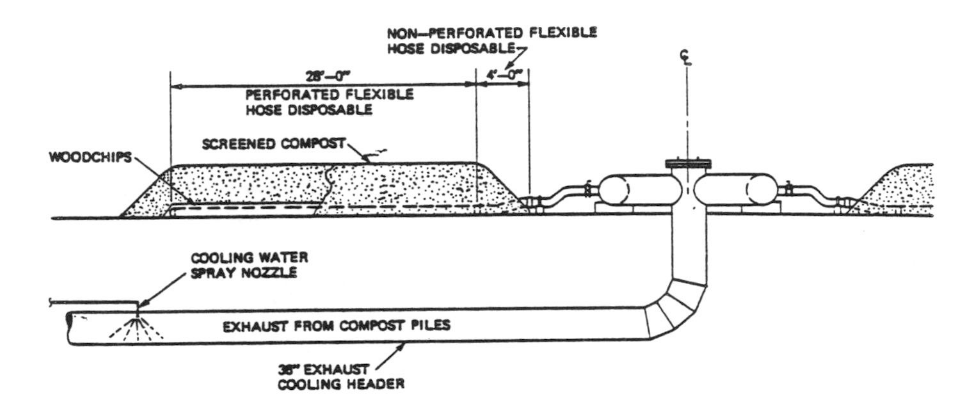

FIGURE 19. SITE II FACILITY ODOR CONTROL FILTER PILE SCHEMATIC

Sample location (cont'd)	Ammonia concentration, ppm
Teardown, 14-day pile, 8 ft in from toe	5
Teardown, 21-day pile, 8 ft in from toe	10
Unscreened, cured compost spread for drying	16
Finished compost storage pile	40

Ammonia was not detected in samples taken 3 and 10 ft away from unscreened compost that was spread for drying.

The Columbus facility composts raw sludge so the potential for odor production is quite high. The operators are aware of this potential and perform certain operational procedures to minimize odor generation. Sludge is mixed with wood chips as soon as possible after delivery. This eliminates the greatest potential source of odor. Mixing was originally performed outside, but a new, enclosed mix building with an odor scrubber was built in 1984. Unfortunately, problems have occurred with this facility and because of this the doors are always open and the scrubber is inoperative. The mixture is shielded from the rain which is beneficial. However, as was described earlier, the uniformity is variable, depending on the operator, and this causes the formation of sludge balls in the pile which can cause odor when the pile is torn down. The aeration system is generally operated in the positive mode, and in this mode the insulation layer of the pile acts to scrub odors. In addition, the positive mode (1) often generates lower temperatures which can reduce odor generation potential, (2) can often reduce the moisture content by distributing airflow more uniformly, and (3) prevents condensate generation.

Ammonia, hydrogen sulfide, methyl and ethyl mercaptans, and sulfur dioxide concentrations were also measured during the on-site investigations at the Columbus facility using a Matheson-Kitigawa gas sampling apparatus. None of these gases were detected at any of the locations. However, participants did identify ammonia during pile teardown. There were several other malodors (unidentifiable) released during pile teardown. The facility has been subjected to odor complaints from neighbors although frequency of complaints has decreased recently.

In general, the assessment of odor control features at the three static pile facilities showed that management of odor is complex. Even with appropriate control techniques, the earthy smell of stabilized compost is often detectable in the immediate vicinity of a composting facility. Based on the investigations performed during the technology evaluation, key design and operational features which can minimize generation or release of objectionable odors were identified as follows:

1. Trucks used to haul dewatered sludge should be covered and cleaned frequently. This is particularly important with raw sludge.

2. Dewatered sludge deliveries should be managed such that mixing and other operations can be performed without sludge accumulating for long periods of time. This is particularly important with raw sludge and in hot weather.

3. An initial mix moisture content of 60 percent or less (total solids ≥40 percent) is a key process design criterion for effective performance, including odor control. Enclosing the mixing operation, or other operations, and scrubbing exhaust gas can be an effective step for odor control in some situations.

4. A uniformly mixed and sufficiently porous material is important for odor control during composting. The presence of clumps of unmixed sludge can lead to anaerobic and, thus odorous, conditions, as well as incomplete stabilization and pathogen inactivation. Construction of daily static pile compartments, with cover material, is also important.

5. Positive aeration during active static pile composting can minimize odor generation potential since the pile cover material acts as an odor scrubber. Negative aeration requires the use of a separate exhaust scrubber system. Regardless of the aeration mode employed, control of aeration rate, oxygen content, and temperature is critical for effective odor control (and proper composting).

6. Teardown of static piles can be managed to minimize release of odors during wet weather or in eartly morning hours when air inversions may result in odor generation. Alternatively, high-rate aeration for 24 to 48 hours prior to teardown in order to reduce pile temperature to ambient conditions can be employed.

7. An effective leachate, condensate, and runoff collection and disposal system will minimize odor generation potential. Proper site drainage is required to prevent ponding which can generate odor.

8. Effective housekeeping procedures such as washing equipment and flushing or sweeping working areas also reduce odor generation potential.

Aerosol Measurement and Control--

There are no dust control facilities at the Hampton Roads facility. The normally wet and moist weather aids in reducing dust problems; however, during dry weather, dust filters on the equipment clog and frequent cleaning is required. Dust accumulation measurements obtained in August 1984, as part of the technology evaluation, indicated the heaviest dust accumulation in the near vicinity of the screening operation under the covered mixing and screening shed. Dust was blown in a southeasterly direction (the prevailing wind), and dust accumulation rapidly declined with distance from the screening operation such that the weight of particulates 300 to 400 ft downwind was only 1 to 3 percent of that immediately adjacent to the operation.

A study performed by the Hampton Roads Sanitation District determined that the Hampton Roads facility is a source of Aspergillus fumigatus aerospora. However, concentrations were not great enough to increase ambient levels in nearby populated areas. Control tests showed that industrial sites, including a municipal landfill, were larger sources of A. fumigatis aerospora than the composting facility. Measurements taken up to 3,300 ft from the facility detected spores only in a nearby maintenance station, located 1,300 ft southeasterly of the facility in the predominant wind direction. Additionally, the spore was only detected during periods of windy weather. Concentrations of A. fumigatis aerospora declined rapidly with distance from the facility. The findings from this study were used as a basis for revising permit monitoring requirements, such that, currently, monitoring for the spore is conducted once per quarter at this facility.

Dust at the Site II facility is controlled by use of covered buildings and an exhaust air filtration system in the screening building. Screens are covered by exhaust hoods to remove particulates generated during the screening process. Table 18 presents data from a study which was performed to determine if workers in the compost screening building must continue to wear respirators. The study was performed for the Site II facility by an independent consultant and included measurement of A. fumigatus aerospora and thermophilic fungi as well as total dust particulates. The study indicated that levels of A. fumigatus aerospora and thermophilic fungi were within the limits of expected background concentrations. Furthermore, dust concentrations at locations in the center of the building were considered to be fairly typical of normal exposure levels. The highest particulate concentration detected, 23 milligrams per cubic meter (mg/cu m), was obtained from monitoring directly under a conveyer where screening was in process, and represents the highest level to which a worker might be exposed. The study concluded that there is no hazard to the workers under normal conditions; however, if an employee were required to work continuously under the screening equipment, precautions should be taken. It was also noted that screening took place only intermittently prior to the testing. Thus, if screening were a continuous activity, additional measurements should be taken to ensure that there are not changes in the exposure levels.

Routine ambient air monitoring at the Site II facility consists of sampling at the composting site and at a nearby off-site location twice a month, plus sampling at 14 other off-site locations once a month. Parameters analyzed include (1) aerobic bacteria, (2) mesophilic fungi, (3) thermophilic fungi, (4) A. fumigatus aerospora, (5) fecal streptococcus, and (6) fecal coliform. Routine monitoring for the aerosols to date has not detected any significant increase in the levels of these organisms based on facility records.

At the time of the on-site investigations at the Columbus facility (winter 1984-85), dust was not an issue; however, accumulation of water on the surface of the site was apparent. The water accumulated primarily around

TABLE 18. SITE II FACILITY AEROSOL MONITORING

Sampling date	Location	Particulate concentration (mg/m^3)	*Aspergillus fumigatus* concentration, cfu/m^3			Thermophilic fungi concentration, cfu/m^3		
			>8 μm	<8 μm	Total	>8 μm	<8 μm	Total
12/16/83	Center of Screening Building	1.6	0.0	0.88	0.88	9.70	8.82	18.52
1/13/84	Center of Screening Building	1.2	0.0	10.5	10.5	93	323	416
1/13/84	Under conveyer during screening	23.0	-	-	-	-	-	-

Note: Based on facility records.

the mixing building, due apparently to settling of the asphalt floor. Columbus has a vehicle to spread water around the site for dust control if this is necessary during the summer months. Since the site is all paved, dust should generally be minimized with proper housekeeping. Additionally, the main screen is now being housed under a roof which should further minimize dust generation potential.

A. fumigatus aerospora is routinely measured at the Columbus facility. The data examined show that the values were quite variable, probably because sampling was performed during various undefined site activities. No levels were unusually high. This does not seem to be a major issue at this compost facility. All loaders are enclosed and use filtered air to protect the operators.

Information on aerosol measurement and control which was obtained at the three static pile facilities during the technology evaluation indicates that dust generation and release of biological agents such as A. fumigatus aerospora have not been a problem. However, site-specific factors such as climate, equipment application, and siting need to be considered where this technology is applied at other locations.

Leachate, Condensate, and Runoff Control--

Runoff at the Hampton Roads facility is drained from all concrete slabs and roadways into a sewer for treatment at the James River treatment plant. The aeration troughs are sloped to collect leachate and condensate. However, no leachate or condensate was observed during the on-site investigation because the aeration system was operated in the positive mode. A torrential rainstorm of approximately 2 hours duration occurred on July 18, 1984, during the on-site investigation. The drainage system was generally effective in removing the rainwater. Some temporary ponding occurred because of clogged manhole grates; however, this was corrected by scraping with a front-end loader or small tractor.

Leachate generated during composting at the Site II facility is collected by surface swales provided for this purpose and eventually drains into a nearby catch basin. Condensate is intercepted before entering the aeration blower, drained from the suction header and also collected by a surface drain discharging into a nearby catch basin. A containment pond is provided at the Site II facility for intercepting stormwater runoff and area washdown waters from the site process areas. Leachate, condensate, and outflow from the containment pond are discharged to a sanitary sewer system. Roof runoff is intercepted and transported to natural drainage swales.

Runoff at the Columbus facility is drained from all concrete slabs and roadways to a leachate lagoon. The water from the lagoon is pumped to the Southerly plant for treatment. There is no condensate since the aeration system is operated in a positive mode. The site apparently does not have an adequate slope to allow runoff to move rapidly off site, since ponding was evident at several locations during the on-site investigation. This could be a potential odor problem and, in addition, much of the ponded water can be absorbed into the composting mass.

POSTCONSTRUCTION COMPARISON

Key features common to current (postconstruction) operations at all of the static pile facilities studied are summarized in Table 19. Key postconstruction facility modifications which have been implemented in response to problems experienced since initial facility construction are presented in Table 20. Both tables are based on information presented in Chapter 3 as well as this chapter. Key findings from the postconstruction comparison of static pile facilities are presented below:

1. All facilities use the extended static pile process to compost either raw or anaerobically digested municipal sludge generated at activated sludge wastewater treatment plants.

2. Trucks are used to haul dewatered sludge from the treatment plants to the composting site. Haul distance varies from 6 to 35 miles.

3. Sludge loadings are variable even though deliveries are coordinated with wastewater treatment plant operation. Peak-to-average day loading ratios vary from 1.4 to 1.9. Higher-than-anticipated ratios can (1) increase mixing and pile construction time, (2) reduce personnel time available for other operations, and (3) utilize active composting area more rapidly than expected. Generally, however, high peak loadings occur for only a limited duration and do not cause sustained operational impacts. Loadings based on operating days more accurately reflect day-to-day materials management and process requirements, although calendar-day loadings are useful for general comparisons.

4. Total solids content of sludge typically received at each facility (17 percent) is below that estimated at design by 3 to 5 percent. As a result, the quantity of wood chips (used as a bulking agent at all sites) required is as much as 80 percent over the design estimates. Typical mix ratios are between 3.5 and 4.5 cu yd/wt, depending on season, proportion of new and recycled chips, and chip moisture content.

5. Increased wood chip use impacts material flow through the composting process, materials storage, equipment mobility, and operating cost. For example, at a fixed-pile height, more active composting area per ton of sludge is required with larger bulking agent quantities. Similarly, storage area for wood chips or unscreened compost, is also greater. Equipment mobility may be constrained if access ways or operating areas are encroached upon due to materials storage constraints. Purchase cost for new wood chips, if required, may strain operating budgets, thus, limiting funds for other facility needs. These problems have been resolved by (1) modifying on-site features and operations to increase materials management capabilities, (2) purchasing additional materials handling equipment, (3) investigating means of improving sludge dewatering; (4) managing operations to maximize the use of recycled wood chips, and (5) modifying operating budgets to purchase more wood chips.

TABLE 19. COMMON FEATURES OF STATIC PILE FACILITIES

Item	Description
General features	
Process	Aerated static pile.
Treatment plant(s)	Activated sludge plants located off-site.
Sludge transport	Truck haul of dewatered sludge.
Sludge loadings	
Variability	Flexibility to respond to variable sludge quantities is a key operating requirement.
Total solids	Dewatered sludge total solids content is typically 3 to 5 percent below design estimates.
Deliveries	Coordinating sludge deliveries is important for materials management and odor control.
Mixing operations	
Site features	Mixing areas are paved and protected from inclement weather.
Equipment	Front-end loaders are used for rough mixing at all facilities. Fine mixing equipment (front-end loaders, mobile composters, a manure spreader, and a rototiller) varies with facility.
Bulking agent	Wood chips are used and are stored in paved, uncovered areas.
Performance criteria	Mix homogeneity and a minimum initial mix total solids content of 40 percent are required.
Mix ratio	Typical mix ratios are 3.5 to 4.5 cu yd/wt at dewatered sludge total solids concentrations normally received. Wood chip use is as much as 80 percent greater than design estimates at these mix ratios.
Mixing time	About 40 to 45 minutes are required to achieve a uniform mix for the sludge quantities, mix ratios and mobile equipment used[a].
Pile construction	
Pile dimensions	Heights of 12 to 13 ft are typical but lengths and widths of daily compartments vary.
Base material	A minimum depth of 12 inches is used, and new or unscreened wood chips are employed.
Cover material	A minimum depth of 18 inches is used, but materials vary.
Active composting	
Process mode	The extended static pile mode is used.
Site features	Concrete pads with or without sunken aeration troughs are used.
Performance criterion	Internal temperatures $\geq$ 55 degrees C for at least 3 days at all monitoring points.

[a]Based on operations at the Hampton Roads and Columbus facilities. Mixing time was not determined for the Site II facility. Representative mix quantities are 42 wet tons (100 cu yd) based on summer operation at the Hampton Roads facility and 86 wet tons (145 cu yd) based on winter operation at the Columbus facility.

[b]Because of drying and screening problems at the Columbus facility these features are based only on the Hampton Roads and Site II facilities.

TABLE 19 (continued)

TABLE 19. COMMON FEATURES OF STATIC PILE FACILITIES (continued)

Item	Description
Aeration rate	Rate is initially controlled by cycling blower operation to meet performance criterion, after which high rate continuous aeration is used to lower temperature and promote drying.
Monitoring	Temperature is monitored at multiple points for process control.
Composting period	A minimum of 21 days is used.
Ambient effects	Time to achieve 55 degree C varies with ambient temperature (and other factors), but was less than 18 days.
Drying operations[b]	
Site features	Paved drying areas are used.
Performance	A minimum unscreened compost total solids content of 50 to 55 percent is needed for efficient screening.
Screening operations[b]	
Site features	Screening areas are paved and protected from inclement weather.
Recoveries	Wood chip recoveries based on unscreened compost volume are typically 65 to 85 percent, although recoveries as high as 90 percent can be achieved under optimum conditions.
Curing operations	
Site features	Uncovered curing areas are used.
Curing period	A minimum of 30 days is used.
Finished compost	
Monitoring	Total solids, nitrogen, phosphorus, selected heavy metals, and Salmonella are monitored routinely. Salmonella has never been detected.
Distribution	Sale to local users.
Environmental features	
Odor	Problems, occasionally severe, periodically occur. All facilities utilize odor control facilities and operating procedures.
Aerosols	Dust generation and release of biological agents such as *Aspergillus fumigatus aerospora* are not a problem.
Sidestreams	Leachate, condensate and site runoff are collected and discharged to local sanitary sewers.
Mobile equipment	
Sludge transport	Trucks are used at haul distances of 6 to 35 miles.
Materials handling	Front-end loaders are used extensively for key activities, such as mixing, pile construction, pile teardown and materials transfer.

[a]Based on operations at the Hampton Roads and Columbus facilities. Mixing time was not determined for the Site II facility. Representative mix quantities are 42 wet tons (100 cu yd) based on summer operation at the Hampton Roads facility and 86 wet tons (145 cu yd) based on winter operation at the Columbus facility.

[b]Because of drying and screening problems at the Columbus facility these features are based only on the Hampton Roads and Site II facilities.

TABLE 20. KEY STATIC PILE POSTCONSTRUCTION FACILITY MODIFICATIONS

Item and facility	Modification
Mixing area	
Hampton Roads	Moved from uncovered, concrete pad to covered shed, open on all sides, originally designed for drying.
Columbus	Fully enclosed mixing building constructed to replace uncovered, paved mixing area.
Bulking agent storage	
Hampton Roads	Area paved to replace unimproved surface.
Columbus	Same as Hampton Roads.
Pile construction	
Hampton Roads	Unscreened compost used as base material under toes of pile. Screened compost used as special cover on face of toes. Replaced perforated flexible, plastic aeration pipe with perforated PVC pipe. Modified aeration pipe perforation configuration (see Table 10 and Figures 16 and 17).
Site II	Unscreened compost used as base material under toes of pile. Modified aeration pipe perforation configuration (see Table 10 and Figures 14 and 15).
Active composting	
Hampton Roads	Positive aeration used instead of negative aeration for some operating conditions.
Columbus	Positive aeration used instead of negative aeration for all operating conditions.
Drying operations	
Hampton Roads	Spreading and rototilling unscreened compost on uncovered, concrete slabs replaces drying by negative aeration of 3-ft-tall windrows in a covered shed.
Site II	Restacking unscreened compost with induced aeration for 4 to 5 days replaces spreading and rototilling.
Columbus	Positive aeration in place following active composting and aerated drying/curing (uncovered) has not been successful on a year round basis. Solar drying facilities have been installed but were not evaluated during the technology evaluation.
Screening operations	
Hampton Roads	Moved from uncovered, concrete pad to covered shed, open on all sides, originally designed for drying.
Site II	Ventilation and filter system for fully enclosed screening building is not required.
Columbus	Operations are hampered because of moisture control difficulties and thus a large (10 ac) uncovered, paved storage area is provided for unscreened compost storage.
Curing operations	
Columbus	Aerated curing on a paved area (uncovered) replaces unaerated curing on an unpaved area (uncovered).
Finished compost storage	
Hampton Roads	Area paved to replace unimproved storage areas.
Columbus	Area paved and three-side shed constructed to replace graded storage area.

TABLE 20 (continued)

TABLE 20. KEY STATIC PILE POSTCONSTRUCTION FACILITY MODIFICATIONS (continued)

Item and facility	Modification
Materials handling	
Columbus	Conveyors installed for some materials transfer operations.
Odor control pile	
Site II	Decommissioned and replaced with chemical scrubber system.
Columbus	Decommissioned because positive aeration is used year round.
Hampton Roads	Used only during winter months with negative aeration.
Access roads	
Hampton Roads	Paved surface replaces gravel surface.

6. Paved surfaces are used at all of the static pile facilities for mixing, active composting, drying, screening, and wood chip and finished compost storage areas. Wood chip and finished compost storage areas at two of the facilities were paved following initial construction because of moisture control problems during wet weather. Paved (aerated) curing areas are also used at two locations. Access roads were paved at one location because of wet-weather problems.

7. Covered mixing areas are used at all static pile facilities studied. Partially enclosed or fully enclosed structures are used at two, and an open-sided shed is used at the third. The covered mixing area at two of the facilities was provided following initial construction because of moisture control problems during wet weather. Covered screening is used at two of the three facilites, and at one of these, screening operations were moved under cover following initial construction because of moisture control problems during wet weather. Screening (drying) problems are common to the third facility, particularly during wet weather, and, thus, a large (10 ac) uncovered, paved area for unscreened compost storage is provided. Uncovered areas are used at all facilities for curing and wood chip storage.

8. A minimum initial mix total solids content of 40 percent, coupled with mix homogeneity, is important for effective static pile composting. These mixing performance criteria are met by mixing 40 to 45 minutes with front-end loaders alone (Columbus facility) or in combination with fine-mix equipment such as a manure spreader or rototiller (Hampton Roads facility). Representative mix quantities are 86 wet tons (145 cu yd) and 42 wet tons (100 cu yd), respectively, at these two facilities. Front-end loader mixing without a separate fine-mix step was not successful at the Hampton Roads facility.

9. Active static pile composting is performed on concrete pads, with or without sunken aeration troughs, at all facilities studied. Active composting operations are covered at one location and uncovered at the other two.

10. Extensive testing following initial construction was required at the Hampton Roads and Site II facilities to arrive at reliable pile construction and aeration configurations for routine operation, whereas conventional technqiues are used at the Columbus facility. Extended static pile heights are typically 12 to 13 ft, but lengths and widths of daily pile compartments vary. Minimum depths of base and cover materials are 12 and 18 inches, respectively. Other pile construction features are site-specific.

11. A minimum active composting period of 21 days is required at all of the static pile facilities, but longer periods are sometimes used. PFRP criteria require that internal temperatures at all monitoring points be maintained at $\geq$55 degrees C for at least 3 days. This is

routinely achieved within 21 days at all facilities, even though the time to meet the PFRP criteria varies with ambient temperature. Temperature is monitored at multiple points at all facilities for process control, but oxygen monitoring is performed at only two.

12. At all of the static pile facilities, aeration rate is controlled by cycling blower operation until the PFRP criteria are met, after which high rate, continuous aeration is used to lower temperature and promote drying. Manual blower control is used at two sites and automatic control is used at the third. Both positive and negative aeration are used, and in two cases, positive aeration is used instead of negative aeration as called for during design. Aeration rates per dry ton vary. Number of blowers per pile compartment, individual blower capacity, total blower capacity on site, and blower capacity per wtpd (based on design loadings) are also site-specific.

13. Material composted by the aerated static pile method often is not dry enough to screen directly, and a separate drying step is employed. A minimum unscreened compost solids content of 50 to 55 percent is generally required for effective screening. All static pile facilities investigated have modified drying operations since initial construction. The Hampton Roads facility replaced negative (induced) aeration of 3-ft-high windrows in a covered shed, as designed, with (seasonal) spreading and rototilling on uncovered, concrete slabs specifically constructed for this purpose. The Site II facility now uses a restacking procedure with induced aeration for 4 to 5 days after active composting in place of spreading and rototilling. Positive aeration in place at the Columbus facility, as designed, has been replaced with solar drying facilities. The latter has only recently been placed in operation and was not investigated during the technology evaluation. As noted in item 12, all facilities also use high-rate, continuous aeration during the final stage of active composting to promote drying.

14. The screening operations assessment at the Hampton Roads, Site II, and Columbus facilities demonstrates the need to define screening effectiveness, including expected variability, in aerated static pile applications. Under optimum conditions, wood chip recoveries equivalent to 80 to 90 percent (based on unscreened compost volume) can be obtained. However, this requires that unscreened compost total solids content be between 50 and 60 percent, depending in part on screen size and design. Exceeding the upper limit can result in losses from excessive dust generation, while not meeting the lower limit inhibits the separation of fines from wood chips and leads to wood chip contamination, reduced screening rates, and often mechanical problems with the screening equipment. Wood chip recoveries of 65 to 85 percent are more typical of sustained routine operation based on experience at the three static pile facilities investigated. Ventilated screening operations in a fully enclosed structure with treatment of ventilated air, as designed, is not routinely used at one site because it has not been necessary.

15. Curing operations at each of the static pile facilities studied is dependent, in large part, on materials management requirements. A minimum curing period of 30 days is provided at all facilities. Unscreened compost is cured at the Columbus and Hampton Roads facilities, whereas screened compost is cured at the Site II facility. An advantage of screened compost curing is that it requires less area than unscreened compost curing. Aerated curing is used at the Columbus and Site II facilities but not at the Hampton Roads facility. At the Columbus facility, aerated curing was added after initial construction as a means of improving drying prior to screening.

16. Finished compost is routinely monitored for total solids, nitrogen, phosphorus, selected heavy metals, and Salmonella at all static pile facilities studied. Salmonella has never been detected and levels of other constituents are site-specific. All facilities sell their product to local users.

17. Front-end loaders are used extensively at all facilities for key activities such as mixing, pile construction, pile teardown, wood chip transfer, finished compost loading, and other materials transfer and storage functions.

18. Odor problems of varying severity have been experienced at all of the static pile facilities studied, even though odor control features are a major aspect of their operation. A centralized water spray and compost filter system for treatment of exhaust gas from composting operations at the Site II facility (Figure 19) has been decommissioned because odorous compounds were not contained. A chemical scrubber system has recently been installed at this facility. Filter piles at the Hampton Roads facility are only used during winter operation when negative aeration for active composting is applied. Positive aeration at this facility (summer operation) and at the Columbus facility (year-round operation) make the use of filter piles redundant.

19. Key design and operational features that can be used to minimize generation or release of objectionable odors include (1) covering and routine cleaning of sludge transport trucks, (2) managing sludge deliveries to prevent sludge accumulating on site, (3) ensuring that initial mix moisture ($\leq$60 percent) and uniformity is achieved, (4) construction of daily static piles, (5) either negative aeration with effective exhaust gas treatment or positive aeration, (6) appropriate aeration to ensure oxygen and temperature control during the initial stages of active composting, (7) high-rate aeration for 24 to 48 hours prior to teardown, (8) managing pile teardown during unfavorable weather conditions, (9) managing site drainage and sidestream collection and disposal, and (10) frequent and effective site cleanup.

20. Dust generation and release of *A. fumigatus aerospora* are not a problem at the three static pile facilities based on information obtained during the technology evaluation. Leachate, condensate, and site runoff are collected and discharged to local sanitary sewers at all facilities studied.

6. Windrow Composting Operations Assessment

This chapter presents an assessment of current operations at two windrow composting facilities investigated during the technology evaluation: (1) the Los Angeles County Sanitation Districts Joint Water Pollution Control Plant Composting Facility (Los Angeles facility) located in Carson, California; and (2) the Metropolitan Denver Sewage Disposal District Number One Demonstration Composting Facility (Denver facility) located in Denver, Colorado. The former is a conventional windrow operation, and the latter, which was constructed for demonstration study purposes, employs an aerated windrow process. The chapter is organized into four main subsections. The first subsection describes the operations assessment methods; the second and third present assessment results for the Los Angeles and Denver facilities, respectively; and the fourth provides a postconstruction comparison of the two facilities.

OPERATIONS ASSESSMENT METHODS

The operations assessment of the Los Angeles facility is based on (1) a 1-day site visit conducted June 15, 1984; (2) a review of key published information (1-9) provided by facility personnel; and (3) discussions with personnel at the site. No independent measurements were taken during the on-site work at this installation.

The operations assessment of the Denver facility is based on a preliminary site visit conducted on June 13-14, 1984, and a follow-up on-site investigation August 6-9, 1984. During these site visits, demonstration study records and planning documents were reviewed, on-site observations were made, and interviews with facility personnel were conducted. To identify generic features of the aerated windrow technology as applied at the Denver facility, an independent review and analysis of demonstration study data was also made. No independent measurements were obtained.

LOS ANGELES ASSESSMENT RESULTS

This subsection contains an assessment of current operations (1984-85) at the Los Angeles facility. The assessment includes a review of sludge loadings and characteristics, windrow process technology, including windrow formation and active composting operations, product disposition, and environmental controls. A description of the Los Angeles facility has been presented previously in Chapter 4.

Sludge Loadings and Characteristics

The Los Angeles facility processes dewatered sludge from the Los Angeles County Sanitation Districts Joint Water Pollution Control Plant. This plant consists of advanced primary sedimentation using anionic polymers and secondary treatment using pure oxygen. Combined primary and secondary sludge solids are anaerobically digested in 37 mesophilic digesters. The average solids residence time in the digesters is 17 days. Cellulose and hemicellulose, which are very rapidly decomposed by aerobic organisms in the composting environment, are only partially degraded under this type and length of anaerobic process, and, thus, significant amounts remain in the digested sludge.

Digested sludge is dewatered by both basket and low-speed scroll-type centrifuges. Forty-four basket centrifuges are in service, and they use 4.5 lb of cationic polymer per dry ton of suspended solids to produce sludge cake at 21 percent total solids. Nineteen low-speed scroll centrifuges are in service, which use 7.0 lb of cationic polymer per dry ton of suspended solids to produce sludge cake having 22 to 25 percent total solids. Total daily cake production from the Joint Water Pollution Control Plant is 1,500 wtpd. Approximately 1,000 wtpd is hauled to a landfill, and approximately 500 wtpd is composted by the Los Angeles facility. Characteristics of dewatered sludge received at the Los Angeles facility are summarized below:

Item	Value, percent
Total solids	22-25
Total volatile solids	50
Nitrogen (as N)	2
Phosphorus (as P_2O_5)	3
Potassium (as K_2O)	0.1
Bulk density, lb/cu yd	1,800

Windrow Formation Operations

To initiate windrow formation operations, semitrailers are first loaded with sludge and amendment (see Chapter 4) where some initial mixing occurs. The trucks then drive to a compost pad and discharge their load. Historically, three types of trailer bodies have been used for sludge transport at the Los Angeles facility. One type has a chain-and-flight conveyor running lengthwise in the bottom of a trapezoidal-shaped body. Frequent breakdowns have occurred with this type of body because the flight conveyor contains many moving parts which did not stand up to the abrasive nature of compost. Also, certain combinations of compost and wet sludge formed a stiff mixture that bridged over the conveyor and prevented discharge. Even when the flights did work, the trucks required a long time (20 minutes) to empty. After several years of use, these trailers were retired from service, and both end-dump trailers and power-ram, horizontal-discharge trailers were purchased. Unloading time for the two types of trailers currently employed is approximately 2 minutes.

Combining sludge and amendment in a semitrailer provides only limited initial mixing, and thus the method of forming a windrow for active composting includes several preliminary steps. First, two rows of sludge-amendment mix are placed side by side as the semitrailer unloads. Each row is then rough mixed with a front-end loader, after which the two rows are pushed into a single windrow. Two front-end loaders are used for amendment loading and these windrow formation operations.[8] One is equipped with a 4.5-cu yd bucket, and the other is equipped with a 5.75-cu yd bucket. After the preliminary rough mix during windrow formation, initial time mixing is performed using a Scarab I mobile composter. The Scarab I composter has the capacity to turn approximately 7 tons per minute of wet mixture having a bulk density of 1,500 to 1,700 lb/cu yd and is powered by a 275-hp diesel engine. Windrows formed in this manner typically are 800 ft long, contain up to 525 wet tons of dewatered sludge, and are 4 to 5 ft high by 14 ft wide at the base. A schematic of windrow formation operations is shown on Figure 20.

Active Composting Operations

The active conventional windrow composting period at the Los Angeles facility varies from 30 to 90 days, depending on ambient temperatures and drying rates. During the initial week of active composting, six windrows such as those described above are formed. Historically, these six windrows were then combined into four intermediate-size windrows and, ultimately, into a single large windrow such as illustrated on Figure 21. However, recently, this procedure has been modified such that three of the small windrows formed during the first week are now combined to form a single large windrow. Each large windrow typically is 800 ft long and 7 ft high by 23 ft wide at the base. The large windrows are torn down at the end of the active composting period, and the finished compost is either delivered to the Kellogg Supply Company for distribution or retained for recycling.

During active composting, the Scarab I mobile composter is used to aerate and mix (by turning) the small windrows, and a larger machine (Scarab II) is used to turn the large windrows. The Scarab II has a capacity of 11 tons per minute based on turning a mixture having a density of 1,500 to 1,700 lb/cu yd. A turning frequency of three times per week provides adequate aeration and enhances uniform temperature development without affecting drying. Table 21 is a summary of windrow properties for each of the mobile composters used at the Los Angeles facility. Note that the large windrows process a greater volume of material per linear foot and per acre than do the small windrows.

Internal temperatures in the large windrows are monitored during active composting to demonstrate compliance with a performance criterion which requires that windrow temperatures $\geq$55 degrees C be maintained for at least 15 days. Personnel at the Los Angeles facility have determined that because of the low surface-to-volume ratio of the large windrows (see Table 21), heat is conserved, thus, promoting temperature development for pathogen inactivation.

[8]This mixing method has not been entirely effective, and pugmill mixing is being considered as a replacement.

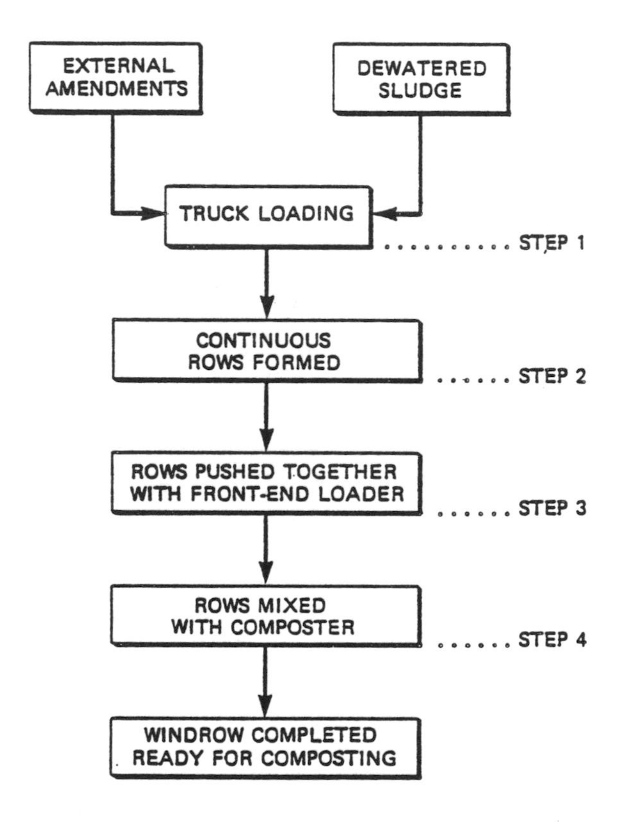

FIGURE 20. SCHEMATIC OF WINDROW FORMATION OPERATIONS AT LOS ANGELES FACILITY (FROM REFERENCE 1)

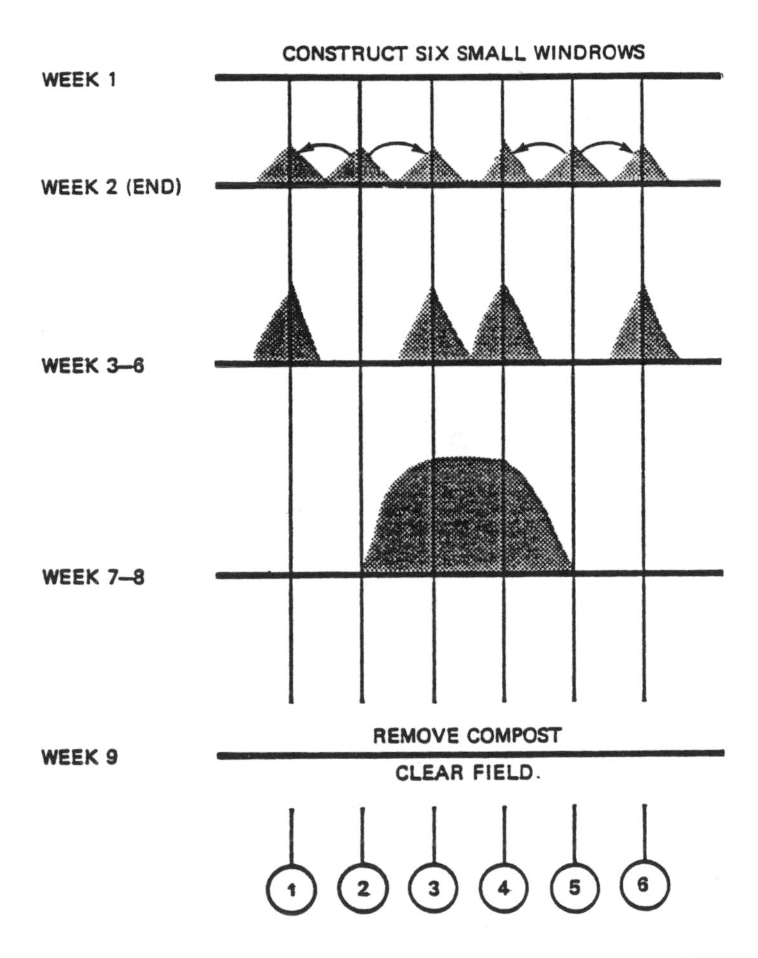

FIGURE 21. ACTIVE WINDROW COMPOSTING OPERATIONS AT LOS ANGELES FACILITY (FROM REFERENCE 1)

TABLE 21. WINDROW PROPERTIES FOR THE LOS ANGELES FACILITY

Mobile composting machine	Windrow property		
	Volume per 100' linear feet, cubic yards	Volume per acre of land, cubic yard/acre	Surface to volume ratio, per ft
Scarab I	125	1,800	0.60
Scarab II	335	3,500	0.32

Both the Scarab I and the Scarab II mobile composters are designed specifically for windrow composting. Each machine straddles the windrow and has a high-speed rotating drum at ground level. The drum has flails which lift the sludge up and over the drum, depositing it behind the machine in windrow form. Turning mixes the wet cake and amendment, increases porosity in the windrow to maintain aerobic conditions, promotes drying of the sludge by exposure to air and sun, and ensures that all of the sludge is subjected to high internal temperatures within the windrows.

Experience at the Los Angeles facility has established that for composting to proceed satisfactorily, windrows should have an initial total solids content of at least 40 percent. This is achieved by mixing dewatered cake with previously finished compost material or other amendment. A total solids content of less than 40 percent results in a mixture of low porosity which inhibits oxygen transfer within the windrow. Initial volatile solids content of windrows using only recycled compost amendment is typically about 45 to 50 percent. When the volatile solids have been reduced to 40 to 45 percent, and the total solids have increased to between 60 and 65 percent, the composting process is completed. After teardown of the large windrow, material is trucked to a stockpile area. After stockpiling, delivery is made to Kellogg Supply Company which screens all material for objects ≥3/8 inch prior to bagging and sale.

Materials Balance

A representative materials balance has been made by personnel of the Los Angeles facility for a 10-dtpd sludge loading using all recycled, finished compost as bulking agent. The balance represents operations with dewatered sludge and recycled compost total solids contents of 23 and 60 percent, respectively. Values are presented below:

Process material	Mass rate, dtpd
Dewatered sludge	10.0
Recycled, finished compost	19.6
Compost mixture	29.6
Composting losses	
Water	28.6
Solids (volatile)	2.6
Finished compost	27.0
Finished compost to stockpile	7.4

The balance shows that on a dry-weight basis, finished compost production for stockpiling is about 74 percent of the incoming sludge loading. Finished compost production is equivalent to 1.9 cu yd per dry ton (0.4 cu yd per wet ton) based on this balance and a finished compost density of 1,300 lb/cu yd.

Product Quality

The general properties of finished compost from the Los Angeles facility are listed below:

Item	Value
Total solids	
Compost with recycled sludge amendment	60-65 percent
Compost with sawdust or rice hull amendment	50-55 percent
Volatile solids	
Compost with recycled sludge amendment	40-45 percent
Bulk densities	
Compost with recycled sludge	1,300 lb/cu yd
Compost with sawdust amendment	875 lb/cu yd
Compost with rice hull amendment	850 lb/cu yd
Carbon to nitrogen ratio	
Compost with recycled sludge	12
Compost with rice hulls	30
Compost with sawdust	135
Particle size	
Median size	1.5 mm
90 percent of particles	less than 5 mm

Table 22 presents a summary of cadmium, lead, and polychlorinated biphenyl (PCB) concentrations detected in compost products from the Los Angeles facility. Heavy metal content of the finished compost has increased since 1977. Data prior to 1977 show that finished compost was less contaminated with heavy metals than at present because only large digested sludge particles were captured with old dewatering equipment used at the treatment plant. Cadmium is associated with smaller digested sludge particles being captured with new dewatering equipment. Data from the period prior to 1977 show that finished compost cadmium content averaged 26 mg/kg, compared to current values of 50 to 70 mg/kg.

Odor Control

The quantity of sludge composted at the Los Angeles facility is limited by odor generation potential (Chapter 4). Odor generation is therefore an important operating parameter. Odor measurements at the Los Angeles facility

TABLE 22. CADMIUM, LEAD, AND PCB CONTENT OF LOS ANGELES WINDROW COMPOST PRODUCTS

Compost product	Year	Concentration, mg/kg dry weight		
		Cadmium	Lead	PCBs
Nitrohumus	1983	60	470	-
	1984	70	510	0.4
Amend	1983	26	250	-
	1984	57	350	0.3
Topper	1983	32	230	-
	1984	47	330	0.3
Recycled compost	1982-83	-	-	1

indicate that typically 83 percent of the windrow odor emissions are the result of ambient surface emissions and 17 percent are the result of windrow turnings. This is equivalent to 30,000 odor units per square foot (ou/sq ft) for ambient surface emissions and 6,000 ou/sq ft for the windrow turnings. These values are based on a 40-day active windrow composting period with 20 turning cycles.

Ambient surface emissions from a windrow decrease significantly as the active compost cycle progresses. This is evident from the chart presented on Figure 22. Emissions are also the greatest immediately after windrow turning, as shown on Figure 23. Experience at the Los Angeles facility has indicated that the best way to control odors and minimize complaints is to limit the size of the composting operation. Thus, the maximum amount of sludge which can be composted in the summer without odor complaints is 500 wtpd. Due to lower productivity in the winter, the annual average is 425 wtpd, on a calendar-day basis.

Ambient surface emissions from windrow composting cannot be eliminated but are subject to some control based on the type of amendment used. Even when composted quantities are within the prescribed limits, certain meteorological conditions (e.g., inversion layers) can cause odor complaints. Various chemical-masking agents have been tried but have been found to be ineffective.

Dust Control

The three major sources of dust at the Los Angeles facility are (1) truck unloading and mixing operations, (2) windrow turning, and (3) surface dust which blows from the windrows and stockpiles. Dust problems are most severe when sawdust is used as the bulking agent.

Dust from truck unloading and initial mixing with front-end loaders is unavoidable under the present operating mode. A proposed, enclosed-mixing facility for improved blending of sludge and amendments would be beneficial in minimizing dust from this operation. Dust blowing from the surface of the windrows and from stockpiled sawdust is handled several ways, as described below.

Surface dust from windrows is most severe in the latter stages of active composting (about 6 weeks into a compost cycle). Allowing the compost to dry only to a point which limits dust generation potential has been effective. Depending upon which compost product is being produced and which amendment(s) is used, the level of dryness varies from about 50 to 55 percent total solids. However, the Kellogg Supply Company prefers a drier finished compost to prevent bags ripping from the additional weight associated with moister compost.

Surface dust from stockpiled sawdust will ultimately be controlled by storage in an enclosed mixing facility which is being planned. Measures now in use to minimize dust generation from this source include spraying the sawdust piles using sprinklers and/or spraying these piles with iron salts

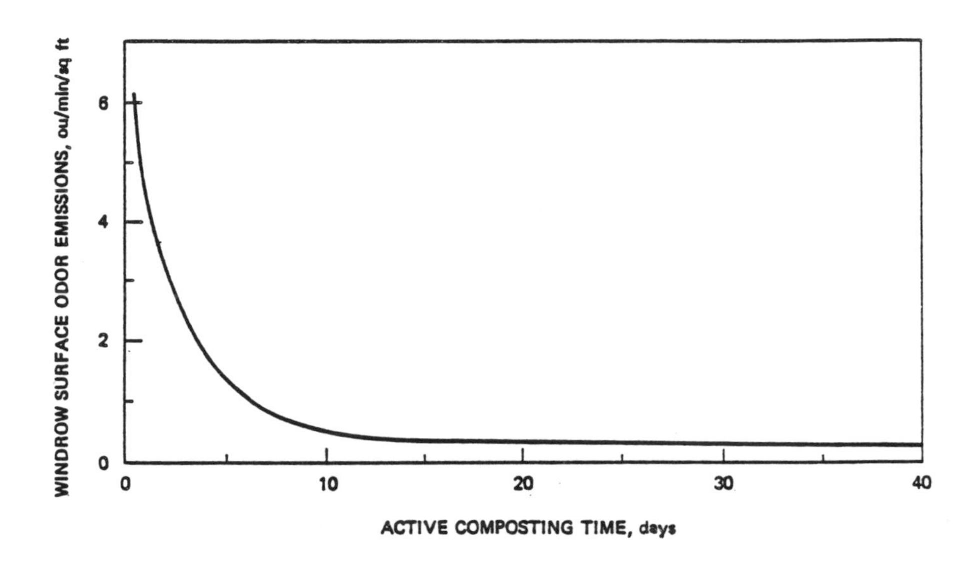

FIGURE 22. AVERAGE WINDROW SURFACE ODOR EMISSIONS DURING A COMPOST CYCLE (FROM REFERENCE 1)

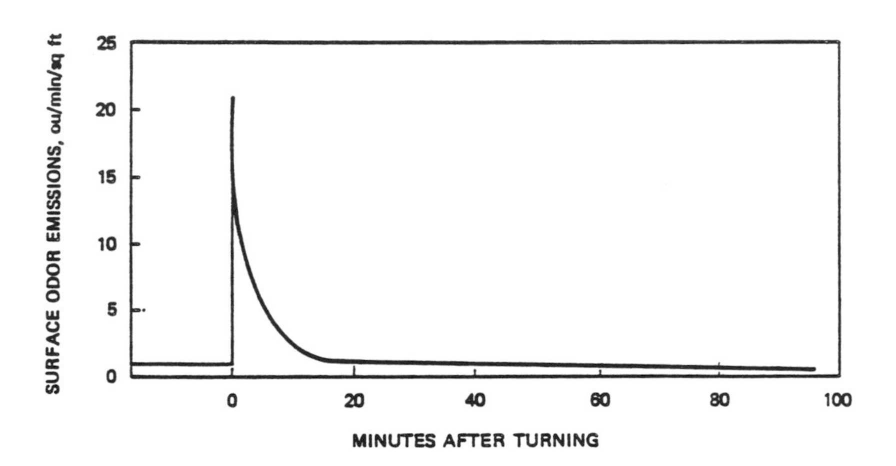

FIGURE 23. SURFACE ODOR EMISSIONS AFTER WINDROW TURNING (FROM REFERENCE 1)

to form a crust on the pile surface. To control dust from windrow turning, this operation is performed only during morning hours (work shift begins at 5:00 a.m.) or before typical daily afternoon breezes begin and is avoided on very windy days. Dust from the asphalt pad and roads is controlled by wetting with a water truck.

Leachate and Runoff Control

The asphalt south composting field (see Chapter 4, Figure 9) is sloped so that any surface runoff flows into a collection sump and then to a holding pond. Overflow from the pond discharges to one of the adjacent treatment plant's influent sewers for treatment. The runoff control system is designed to handle a 50-year storm.

No leachate generation has been detected from the windrows which use sawdust as the bulking agent. However, leachate has been detected from the windrows which use rice hulls. The leachate caused severe deterioration of unsealed asphalt in the active composting area, which was ultimately corrected by applying a coal tar pitch-sand slurry sealer to new and deteriorated surfaces.

Aspergillus Fumigatus Aerospora Impacts

Air sampling by personnel at the Los Angeles facility has shown that *Aspergillus fumigatus aerospora* near the facility are not a problem. The facility reports that from their studies the geometric mean of all upwind samples at the plant was 1.2 colony forming units per cubic meter (cfu/cu m), while that for downwind was 2.1 cfu/cu m. A typical average for outdoors as reported in the literature is 3.2 cfu/cu m.

Wet-Weather Impacts

Historically, wet weather has significantly impacted operations at the Los Angeles facility. Minor rainfall is not a problem. However, intense storms do cause operations to be suspended. Use of an asphalt-paved composting pad has increased winter productivity over the lime-stabilized earthen pad initially employed. Landfilling, the alternative disposal method, is also adversely affected by severe wet weather and must be curtailed during these periods. As a result, all sludge must be stored during severe wet-weather periods. The Joint Water Pollution Control Plant sludge storage silos have a combined capacity of 6,600 wet tons, which is roughly 4.4 days of sludge production. However, due to normal, dry-weather overnight and weekend storage requirements, only 3,600 wet tons of storage capacity (2.4 days of sludge production) is generally available in the wintertime to handle sludge storage requirements caused by adverse weather.

Product Disposition

Substantially all compost product from the Los Angeles facility is sold to the Kellogg Supply Company which has been independently marketing sewage sludge products since the 1920s. In the fall of 1982, bids were requested from other fertilizer manufacturers for the rights to the composted sludge. Every known company in the western United States was contacted and invited to bid; however, only Kellogg Supply Company submitted a bid. A 15-year contract was signed with the company in December 1982.

The contract calls for Kellogg Supply Company to market all of the first 40,000 dry tons of sludge cake composted per year. The company has an option to purchase any additional composted sludge cake produced beyond the first 40,000 dry tons each year. The Los Angeles facility has the right to use 100 dry tons per year of finished compost for general landscaping needs on site. The maximum amount of composted sludge cake taken by Kellogg Supply Company in any single year was 44,000 dry tons in 1977. This was a year of severe drought in California, and very little rainfall occurred to impact the composting operation.

Revenue generated from the sale of finished compost to Kellogg Supply Company is a function of the percentage of gross sales of the company. Revenues range from 3.3 percent if gross sales are less than $1 million per year to 1.1 percent if gross sales are in excess of $3 million per year. Therefore, there is an incentive for the company to market more compost and at higher prices.

Kellogg Supply Company sells a range of sludge-based products; however, as presented in Chapter 4, four products (Nitrohumus, Topper, Amend, and Gromulch) make up the bulk of the sales. These products are marketed in over 2,000 retail stores throughout California, Arizona, and the Las Vegas area of Nevada. The natural organic nature of the products and the benefits of using natural organics over chemicals are stressed. No direct advertising to the general public is used, although the company helps pay for newspaper fliers which advertise sales at major retail stores. The company does advertise to the nursery trade through specialized trade publications.

A key to the acceptance of composting at the Los Angeles facility is a public relations program which has built and maintains credibility with the community. The public relations program was initiated in 1978 as a response to the large number of odor complaints that were being received. Prior to initiation of this program, the community did not understand the plant operation and was more hostile than is now the case. A 30-minute information presentation using 35-mm slides was developed, and community meetings with homeowners were held in the evenings to present and discuss plant operations. Volunteers were requested for an advisory committee to act as liaison between the plant and the community. A permanent citizens' advisory committee was formed which provides for representatives from 17 areas around the facility. Each area has about the same number of homes. Fourteen of the seventeen areas chose to participate and selected a representative for the committee. Annual tours of the plant are also conducted for neighbors. A quarterly newsletter keeps homeowners up to date on current activities. Citizens are urged to submit odor complaints directly to plant staff, and the staff are trained in proper conduct to receive these complaints.

DENVER ASSESSMENT RESULTS

Sludge loadings and characteristics, process technology, environmental controls, and an independent analysis of demonstration study results for the Denver facility are presented in this subsection. A description of the Denver facility has been presented previously in Chapter 4.

Sludge Loadings and Characteristics

As noted in Chapter 4, the Denver facility was designed for demonstration study purposes to process 10 dtpd of anaerobically digested sludge from the Central and Northside wastewater treatment plants. The Northside plant has only primary facilities. Digested sludge from this facility is transported to the Central plant where it is combined with Central plant sludge and redigested and/or dewatered. Because actual loadings have been determined by demonstration study requirements and are, therefore, not indicative of a facility operating on a routine basis, loading characteristics are not considered in this report. However, for reference purposes representative metal and nutrient levels of dewatered sludge processed at the Denver facility are presented in Table 23.

Windrow Formation Operations

Typical windrow formation operations at the Denver facility were observed during the technology evaluation. To form a windrow, a quantity of finished compost from an existing pile is removed as finished compost by a front-end loader. The remaining finished compost in the existing pile is then redistributed along the compost pad using a Brown Bear. External amendment material, such as wood chips, if used, is then deposited on the finished compost, and the amendment base is smoothed using the front-end loader. Sludge is deposited directly on the base material by trucks equipped with power-ram, horizontal-discharge trailers. Initial rough mixing of the sludge and amendment is accomplished using five passes of the Brown Bear. Final mixing and shaping is performed with four passes of a Scarab mobile composter. Each windrow that is formed is about 230 ft long with a triangular cross section 13 ft at the base and 4 to 5 ft high.

Operating personnel at the Denver facility consider that the two-step mixing operation is essential for proper mixing and composting. The Brown Bear alone does not provide a uniformly fine mix for active compositng, and, conversely, the mobile composter alone cannot do an effective job of forming a windrow.

Each windrow formed during the demonstration study typically contained about 200 to 250 wet tons of sludge-amendment mix. Reported initial bulk densities varied from about 700 to 1,500 lb/cu yd, based on 75 measurements, with 1,000 lb/cu yd being a representative value. Each windrow has a surface-to-volume ratio of about 0.54 per ft and a volume of about 100 to 120 cu yd per 100 linear feet. Based on 10- to 12-ft acccess ways between windrows, windrow volume per acre is about 1,800 to 2,000 cu yd.

TABLE 23. SLUDGE CHARACTERISTICS FOR DENVER FACILITY

Constituent	Concentration, dry weight basis
Solids, percent[a]	
Total	16
Total volatile	60
Bulk density, lb/cu yd[a]	1,570
Trace minerals, mg/kg[b]	
Cadmium	33-36
Chromium	700-800
Copper	900
Lead	300-500
Nickel	100-200
Zinc	400-1,450
Nitrogen, percent[b]	
Total Kjeldahl	6.4-11.0
Ammonia	5.8
Total phosphorus, percent[b]	2.7
Potassium, percent[b]	0.3

[a]Independent testing during technology evaluation.
[b]From Reference 10.

A time-and-motion study of windrow formation was performed as part of the on-site investigation conducted during the technology evaluation. The results of this study are presented below:

Activity[a]	Duration, minutes
1. Finished compost from existing windrow is redistributed using the Brown Bear	5
2. External amendment is placed using the front-end loader	23
3. Base is smoothed using the front-end loader	2
4. Area is cleaned up using the front-end loader	6
5. Initial rough mixing is performed using five passes with the Brown Bear	30
6. Windrow area is cleaned up using the front-end loader	13
7. Final mixing is performed using four passes with the Scarab mobile composter	55
8. Final cleanup is performed using the front-end loader	6
9. Total windrow formation	140

[a]Sludge delivery time is not included in these activities.

The time-and-motion study shows that about 2 to 2 1/2 hours are required for windrow formation, excluding sludge delivery time. The pile formed during the technology evaluation was comprised of 87.2 wet tons of sludge at 17 percent total solids (15 dry tons), 3.6 wet tons of wood chips (2.5 dry tons), and 145 wet tons of recycled finished compost (45 dry tons). Initial total solids content of the windrow was 26.5 percent. According to facility personnel, operational experience has shown that windrow formation time is usually shorter than that observed during the on-site investigation.

Active Composting

Active composting is typically carried out over a 30-day period, although longer times were sometimes required during the demonstration study. During active composting, the windrows are turned with the Scarab mobile composter, typically two to three times per week. The Scarab requires about 4 1/2 minutes to turn a 230-ft windrow.

The aeration system used at the Denver facility consists of a blower, a manifold system, and an aeration trough capable of drawing air down through the windrow or blowing air up through the windrow. The blower system consists of 10- and 25-hp blowers. Some windrows use one 10-hp blower which is dedicated to one aeration trough, while in other cases, one 25-hp blower supplies multiple troughs. The manifold is a buried solid-metal pipe which leads from a 12-inch outlet in the blower system to the aeration trough. The manifold system is controlled by a series of butterfly valves.

The aeration troughs were formed as part of the compost pad. Each trough is 16.8 inches deep by 9.6 inches across at the base; there is a lip 2 1/2 inches from the top of the trench and the lip opens up to 12.4 inches at the floor. A cover plate fits into the lip; the cover plate has 0.3-inch holes placed on 2.2-inch centers on both edges of the plate. The cover plates were covered with a construction fabric during the demonstration study, which was in turn covered with a fine sand. Aeration troughs run the length of the compost rows, and there are stainless steel gates dividing the windrows into quarters.

During the demonstration study, the troughs worked quite well as a fixed aeration system component, although the construction fabric clogged, restricting air movement through the trough. At the time of the on-site investigation for the technology evaluation, a new plastic media was being evaluated. Facility personnel report that use of this media in place of the original media has eliminated plugging. Another problem encountered with this aeration system was that front-end loader tires displaced the fine sand if the loader drove longitudinally along the trough.

Demonstration Study Assessment

The aerated windrow demonstration study was designed to investigate, in part, three key process factors: (1) induced aeration rate and mode, (2) covered versus uncovered operation, and (3) amendment requirements. All variables which affect composting could not be rigorously controlled during the study, and, thus, mean values of various process variables were calculated to provide a generic assessment of the aerated windrow process for the technology evaluation. Study data for the assessment were obtained from records provided by facility personnel.

Table 24 summarizes total solids and total volatile solids changes during covered and uncovered windrow composting at the Denver facility, with and without induced aeration. The active windrow composting period varied from 29 to 43 days, with mean values between 30 and 33 days for the data sets presented. The summary shows that induced aeration improved drying by about 2 to 3 percent based on mean moisture reduction. Note, however, that the initial total solids content of many windrows was below 40 percent, and that mean volatile solids destruction was marginal except for the uncovered windrows employing induced aeration.

TABLE 24. EFFECT OF COVERING ON WINDROW COMPOSTING AT DENVER FACILITY

Composting mode	Amendments	Composting parameter, percent	Stage of compost cycle[a]	Covered windrows		Uncovered windrows	
				Mean	Range	Mean	Range
No induced aeration with turning every other day[b]	Recycled compost with various amendments	Total solids	Initial	31.0	26.8-37.4	30.8	26.8-37.4
			Final	37.8	29.6-47.9	38.2	33.8-48.4
		Total volatile solids	Initial	61.3	45.0-70.8	61.3	45.0-70.8
			Final	61.0	48.7-78.2	60.0	47.8-76.4
		Moisture reduction	-	6.8	-	7.4	-
		Volatile solids reduction	-	0.3	-	1.3	-
Induced aeration with turning every third day[c]	Recycled compost with various amendments	Total solids	Initial	40.1	37.5-42.7	36.1	32.6-38.7
			Final	50.2	45.1-59.4	45.4	41.0-55.8
		Total volatile solids	Initial	59.9	58.7-61.8	63.9	60.5-65.9
			Final	61.1	52.5-72.6	58.0	53.7-62.0
		Moisture reduction	-	10.1	-	9.3	-
		Volatile solids reduction	-	-	-	5.9	-

[a]Initial--sample of windrow material obtained on day windrow was constructed. Final--sample of windrow material obtained on day windrow was torn down.

[b]Six observations.

[c]Nine to eleven observations.

Mean moisture reductions presented in Table 24 were similar for both covered and uncovered windrows, indicating that under the conditions studied covering the windrows did not enhance drying. Additionally, it was observed that in many cases, maximum temperatures for uncovered windrows were greater than 55 degrees C for a longer duration than for covered windrows. The reason for this was not evident from the work done during the technology evaluation.

The effect of amendment use on aerated windrow composting at the Denver facility was assessed based on an analysis of demonstration study data from uncovered windrows which were formed using either recycled compost alone or in conjunction with an external amendment such as wood chips or sawdust. The results are presented in Table 25. For the two sets of data shown, the active windrow composting period varied from 21 to 45 days, with mean values of 29 and 32 days. Eighteen piles which used only recycled compost as an amendment yielded mean initial and final total solids of about 40 and 52 percent, respectively, with corresponding mean volatile solids of 54 and 51 percent. Eleven piles which incorporated other amendments with the recycled compost yielded mean initial and final total solids of 40 and 57 percent, respectively, with corresponding mean volatile solids of 59 and 58 percent. Although mean volatile solids destruction for those windrows which incorporated an external amendment was lower than for those that did not, drying was greater by about 5 percent for the former. Thus, for the conditions investigated, use of external amendment had a beneficial effect on drying during aerated windrow composting.

Finished Compost Quality

Independent measurement of finished compost quality at the Denver facility was not performed during the technology evaluation. Table 26 presents the results of chemical and bacteriological analyses on finished compost samples from selected piles based on plant records. Variations in some of the constituents may be due to the dilution effect exerted by the bulking agents tested.

Fecal coliform and fecal streptococci levels of 10^2 to 10^5 and 10^4 to 10^8 per gram, respectively, were detected in the finished compost. Salmonella levels per gram of 6 or below were also detected. Measurement of *Aspergillus fumigatus aerospora* yielded concentrations in the range of 10^4 to 10^5 colony forming units (cfu) per gram. Facility personnel report that the principal goal of the demonstration study was to study composting and produce a dried product rather than to produce a finished compost meeting specific quality requirements.

Odor Studies

Denver facility personnel routinely monitor odor at two locations on the periphery of the composting site. Three months of odor monitoring records were reviewed as part of the technology evaluation, and the results are summarized below:

TABLE 25. EFFECT OF AMENDMENT USE ON AERATED WINDROW COMPOSTING AT THE DENVER FACILITY

Composting parameter	Stage of compost cycle[a]	Windrows with recycled compost amendment only[b]		Windrows with recycled compost plus external amendment[c]	
		Mean	Range	Mean	Range
Total solids, percent	Initial	39.5	32.0-44.8	39.8	33.5-46.2
	Final	51.8	43.7-57.9	56.8	46.2-65.1
Total volatile solids, percent	Initial	53.8	48.5-63.4	58.9	53.9-63.8
	Final	50.5	44.7-58.8	57.5	46.8-80.5
Moisture reduction	-	12.3	-	17.0	-
Volatile solids reduction	-	3.3	-	1.4	-
Cycle length, days	-	32	21-45	29	27-37

[a]Initial--Sample of windrow material obtained on day windrow was constructed.
Final--Sample of windrow material obtained on day windrow was torn down.
[b]Number of observations: Initial values = 18. Final values = 17. Cycle length = 18.
[c]Number of observations = 11 for all data tabulated.

TABLE 26. FINISHED COMPOST CHARACTERISTICS FOR THE DENVER FACILITY

Constituent	Number of observations	Concentration, dry weight basis
Fecal coliform per gram	9	1.6×10^2 to 1.1×10^5
Fecal streptococci per gram	12	4.4×10^4 to 2.8×10^8
Salmonella per gram	12	1-6
Aspergillus, cfu/gram	12	2.9×10^4 to 9.1×10^5
Chemical species, mg/kg		
Total Kjeldahl nitrogen, as N	10	5,070-16,000
Ammonia nitrogen, as N	11	1,440-3,880
Total phosphorus, as P	6	10,400-13,300
Calcium	6	17,100-19,300
Magnesium	6	2,080-2,330
Potassium	6	2,200-3,250
Sodium	6	620-1,200

Item	Location 1	Location 2
Number of days sampled	22	21
Number of days odor at site boundary required levels of dilution indicated below:		
0	15	14
2	2	4
7	4	1
15	1	2

State of Colorado regulations specify that odor at the boundary be at a level which requires 15 dilutions or less. Thus, for the 3-month period studied, odor at the site boundary was at a level which required the maximum dilution only 1 to 2 days out of the 21 to 22 days sampled.

A special study was performed by the facility to evaluate the effectiveness of various odor scrubber systems. The results, shown in Table 27, were obtained using wood chips, activated carbon, and compost as scrubber media. Wherever possible, two readings were averaged to give the numbers shown. These results indicate that activated carbon is very effective for a short period of time. The wood chip scrubber was quite variable and was rapidly overloaded when two blowers were on at the same time. The compost scrubber was quite effective with one blower over the length of the trial.

Aspergillus Fumigatus Aerospora Impacts

A study of *Aspergillus fumigatus aerospora* impacts was performed at the Denver facility in 1983. Key results from the study are presented below:

1. Agar plate counts downwind from the front-end loader during compost stockpiling operations were about ten times those from upwind plates, indicating that compost operations can be a source of these organisms.

2. On August 11, 1983, quantitative background sampling was conducted around the entire compost site when there was no activity such as stockpiling or windrow turning on the site. The data indicated that there was a very low concentration of *A. fumigatus aerospora* in air samples on the Denver facility compost site in the absence of stockpiling or windrow turning. The highest concentration (14/cu m) was from a location near a blower operating in a suction mode on three windrows. This indicated that the blower operation (suction mode) could be a possible continuous source of *A. fumigatus aerospora*. All spores recovered in the microbial air sampler were in the respirable size fraction.

3. On August 17, 1983, quantitative air sampling was conducted during windrow turning using the Scarab mobile composter. Spores recovered from the microbial air sampler were primarily in the respirable

TABLE 27. RESULTS OF ODOR SCRUBBING STUDY

	Odor level after scrubbing, as percent of unscrubbed exhaust air odor level[a]				
		Finished compost scrubber		Wood chip scrubber	
Days from start	Activated carbon column	1 blower operating	3 blowers operating	1 blower operating	2 blowers operating[b]
5	0.6	0.3	8.9	4.3	48.6
10	28.6	0.6	8.9	8.6	100.0
15	100[c]	0.3	4.9	2.0	9.2
20	-	0.9	2.9	9.2	100.0
25	-	2.6	8.6	9.2	48.6

[a]Odor level for unscrubbed exhaust air equals 350 dilutions.
[b]Highly variable.
[c]Scrubber clogged and failed after 13 days.

size fraction. Airborne spore concentrations were higher than the background values obtained on August 11, 1983. The mean background value was 4/cu m while the mean value during the turning operation was 160/cu m.

4. During the August 17, 1983, sampling the highest concentration of A. fumigatus aerospora recovered was 680/cu m. This occurred while sampling alongside the Scarab mobile composter on the downwind side as it turned the windrow. This allowed the samples to remain in the discharge "plume" during the process of turning the entire length of a windrow. These data indicate that the process of turning compost windrows with a mobile composter disseminates a significant number of spores of A. fumigatus into the surrounding air. The data also indicate that a person close to the downwind side of such an operating machine is likely to receive a higher dose of spores than if they were farther away.

POSTCONSTRUCTION COMPARISON

A comparison of key features of current (postconstruction) operations at each windrow facility studied is presented in Table 28. The comparison is based on information from Chapter 4 as well as this chapter. Although the postconstruction comparison is limited because the Denver facility was built for demonstration study purposes, key findings based on investigations at one or both facilities are presented below:

1. Both windrow facilities compost anaerobically digested sludge generated at activated sludge wastewater treatment plants. The conventional windrow process is used at the Los Angeles facility, and aerated and conventional windrow processes are used at the Denver facility.

2. At both windrow facilities, trucks are used to haul dewatered sludge from the treatment plant to the adjacent composting facility. Sludge storage silos are used at the Los Angeles facility to manage composting loadings.

3. Experience at the Los Angeles facility has demonstrated the need to limit the sludge quantity processed in order to minimize odor complaints at this location. The summer maximum is limited to 500 wtpd, but, because of lower productivities during the winter, the annual average sludge quantity composted is about 425 wtpd. The lower winter productivities are caused by wet weather which has historically impacted operations. Although minor rainfall is not a problem, intense storms cause operations to be suspended.

4. Dewatered sludge total solids content at the Los Angeles facility is 22 to 25 percent, although historically (Chapter 4) lower concentrations were typical. Major improvements to sludge dewatering operations were required to achieve the current levels. The Denver facility typically operates at dewatered sludge total solids concentrations of 16 to 17 percent.

TABLE 28. POSTCONSTRUCTION COMPARISON OF WINDROW FACILITIES

Item	Los Angeles facility	Denver facility
General features		
Process	Windrow	Windrow
Treatment plant	Activated sludge adjacent to site.	Activated sludge adjacent to site.
Sludge transport	Truck haul of dewatered sludge.	Truck haul of dewatered sludge.
Sludge loadings		
Variability	Summer maximum limited to prevent odor. Winter maximum limited by wet weather. Dewatered sludge storage silos used.	
Total solids, percent	22 to 25	16 to 17[b]
Windrow formation		
Equipment[c]	Rough mix with front-end loader followed by fine mix with mobile composter.	Rough mix with Brown Bear followed by fine mix with mobile composter.
Amendments	Recycled compost, sawdust, rice hulls[d]	Recycled compost, wood chips, sawdust, straw, cornstalks, bark chips, tree trimmings[e]
Performance criteria	Mix homogeneity and a minimum total solids content of 40 percent are required for effective composting.	Mix homogeneity and a minimum total solids content of 40 percent are required for effective composting.
Mix ratio	Typically 1.2 cu yd/wt using only recycled compost.[f]	Varied during demonstration study.
Mixing time	-[a]	85 minutes[g]
Dimensions	Small windrows: 4-5 ft high, 14 ft wide at base and 800 ft long. Large windrows: 7 ft high, 23 ft wide at base and 800 ft long.	4-5 ft high, 13 ft wide at base and 230 ft long.
Quantities[h]	Small windrows: 500 wet tons of sludge and 400 wet tons of recycled compost. Large windrows: 3 times values for small windrows.	200-250 wet tons of material per windrow.

[a]Not determined during technology evaluation.
[b]Based on samples obtained during technology evaluation.
[c]Mixing only.
[d]Main amendments used for finished compost products, Nitrohumus, Topper and Amend.
[e]Various amendments were investigated during demonstration study. See Table 5.
[f]Based on dewatered sludge total solids equal to 23 percent, recycled compost total solids equal to 60 percent, and recycled compost bulk density equal to 1,300 lb/cu yd.
[g]Based on time and motion observations during the technology evaluation. Includes initial rough mixing of recycled compost, external amendment and dewatered sludge using a Brown Bear followed by fine mixing using a mobile composter.
[h]Representative values for windrow dimensions noted.
[i]Includes drying in place.
[j]An independent analysis of demonstration study results was performed during the technology evaluation at this facility.
[k]Criteria are for operations using only recycled compost amendment.
[l]See Table 25 for actual performance based on demonstration study results.

TABLE 28 (continued)

TABLE 28. POSTCONSTRUCTION COMPARISON OF WINDROW FACILITIES (continued)

Item	Los Angeles facility	Denver facility
Properties	Small windrows: 125 cu yd per 100 linear feet, 1,800 cu yd/ac, and surface-to-volume ratio of 0.60 per ft. Large windrows: 335 cu yd per 100 linear feet, 3,500 cu yd/ac, and surface-to-volume ratio of 0.32 per ft.	100-120 cu yd per 100 linear feet, 1,800 to 2,000 cu yd/ac, and surface-to-volume ratio 0.54 per ft.
Active composting[i]		
Process mode	Conventional windrow.	Aerated and conventional windrow[j].
Site features	Asphalt-paved pad.	Asphalt-paved site. Sunken aeration troughs for aerated windrow process.
Equipment	Mobile composters rated at 7 and 11 tons per minute.	Mobile composter rated at 5 tons per minute. Induced aeration equipment for aerated windrow process includes 10- and 25-hp blowers.
Performance criteria	Internal temperature ≥ 55 degrees C for ≥ 15 days at all monitoring points. Final volatile solids of 40-45 percent (≥ 5 percent reduction) and final total solids of 60-65 percent.[k]	Internal temperature ≥ 55 degrees C for ≥ 15 days. Minimum final total solids of 60 percent using only recycled compost amendment.[l]
Turning frequency	3 times per week.	2-3 times per week.
Composting period	30-90 days.	30-45 days.
Monitoring	Temperature, for process performance.	Temperature, for process performance.
Finished compost		
Monitoring	See text.	See text.
Distribution	Sale to private company for distribution.	Not applicable.

[a]Not determined during technology evaluation.
[b]Based on samples obtained during technology evaluation.
[c]Mixing only.
[d]Main amendments used for finished compost products, Nitrohumus, Topper and Amend.
[e]Various amendments were investigated during demonstration study. See Table 5.
[f]Based on dewatered sludge total solids equal to 23 percent, recycled compost total solids equal to 60 percent, and recycled compost bulk density equal to 1,300 lb/cu yd.
[g]Based on time and motion observations during the technology evaluation. Includes initial rough mixing of recycled compost, external amendment and dewatered sludge using a Brown Bear followed by fine mixing using a mobile composter.
[h]Representative values for windrow dimensions noted.
[i]Includes drying in place.
[j]An independent analysis of demonstration study results was performed during the technology evaluation at this facility.
[k]Criteria are for operations using only recycled compost amendment.
[l]See Table 25 for actual performance based on demonstration study results.

TABLE 28 (continued)

TABLE 28. POSTCONSTRUCTION COMPARISON OF WINDROW FACILITIES (continued)

Item	Los Angeles facility	Denver facility
Environmental features		
Odor	Limits sludge composting rate. Emissions vary, with composting cycle and turning frequency. Occasional problems particularly during temperature inversions.	Emissions infrequently exceed dilution criteria at site boundary.
Aerosols	Dust can be a problem from truck unloading, mixing, windrow turning, and windrow surfaces. *A. fumigatus aerospora* releases are not a problem.	Compost stock piling, windrow turning and aeration blowers during negative mode are sources of *A. fumigatus aerospora*, but off-site impact was not assessed.
Sidestreams	Sidestreams collected and discharged to sanitary sewer.	Sidestreams collected and discharged to sanitary sewer.

[a]Not determined during technology evaluation.
[b]Based on samples obtained during technology evaluation.
[c]Mixing only.
[d]Main amendments used for finished compost products, Nitrohumus, Topper and Amend.
[e]Various amendments were investigated during demonstration study. See Table 5.
[f]Based on dewatered sludge total solids equal to 23 percent, recycled compost total solids equal to 60 percent, and recycled compost bulk density equal to 1,300 lb/cu yd.
[g]Based on time and motion observations during the technology evaluation. Includes initial rough mixing of recycled compost, external amendment and dewatered sludge using a Brown Bear followed by fine mixing using a mobile composter.
[h]Representative values for windrow dimensions noted.
[i]Includes drying in place.
[j]An independent analysis of demonstration study results was performed during the technology evaluation at this facility.
[k]Criteria are for operations using only recycled compost amendment.
[l]See Table 25 for actual performance based on demonstration study results.

5. Both windrow facilities use two-step mixing as part of windrow formation. The Los Angeles facility uses front-end loaders to rough mix sludge and amendment, whereas the Denver facility uses a Brown Bear. Front-end loader mixing has not been entirely effective at Los Angeles, and pugmill mixing is being considered as a replacement. Fine mixing at both sites is achieved with mobile composters. Recycled compost is the main amendment used at both facilities. Sawdust and rice hulls are also used at the Los Angeles facility, and during the demonstration study at the Denver facility, use of several amendments was studied with varying success.

6. A minimum mix total solids content of 40 percent and mix homogeneity are key criteria for effective windrow composting at both facilities. A mix ratio of 1.2 cu yd of amendment per wet ton of dewatered sludge is typical at the Los Angeles facility when only recycled compost is used. Mix ratios varied during the demonstration study at the Denver facility, and in several instances, the initial mix total solids criterion was not achieved prior to composting trials.

7. Conventional-sized windrows having similar dimensional properties (except length) are used at both the Denver and Los Angeles facilities. Sludge-amendment mix quantity per windrow varies because lengths vary, as does bulk density. Bulk density is affected, in part, by the type of amendment used. During active composting at the Los Angeles facility, large windrows are also constructed from three of the conventional-sized windrows. These large windrows conserve heat and occupy less area per volume of sludge-amendment mix than the conventional-sized windrows.

8. Active composting sites are paved at both windrow facilities, and at Denver, sunken aeration troughs are also used for induced aeration during aerated windrow composting. Mobile composters mix and aerate the windrows during conventional windrow composting at both facilities and during aerated windrow composting at Denver. A turning frequency of three times per week is typical during conventional windrowing and two times per week during aerated windrowing.

9. A minimum active composting period of 30 days is required at both windrow facilities, but longer periods are sometimes used. PFRP criteria require that internal temperatures at all monitoring points be maintained at $\geq$55 degrees C for at least 15 days. Use of large windrows (with low surface-to-volume ratios) during the latter stage of composting at the Los Angeles facility conserve heat and aid temperature development. Based on demonstration study results, temperature development during conventional and aerated windrow composting at the Denver facility was varied. Temperature is monitored at multiple points at both facilities.

10. At the Los Angeles facility, conventional windrow composting with recycled compost amendment is considered complete when windrow total solids content reaches 60 percent and volatile solids are reduced to

40 to 45 percent. A volatile solids reduction of about 5 percent is typical. When rice hulls or sawdust are used in conjunction with recycled compost, the final total solids content of windrows is typically 50 to 55 percent. Although a final total solids content of 60 percent was established as a performance criterion for aerated and conventional windrow composting during the demonstration study at the Denver facility, this was difficult to achieve under the conditions investigated.

11. Based on an independent analysis of demonstration study results from the Denver facility, induced aeration improved drying by about 2 to 3 percent over conventional windrowing. Use of external amendments in conjunction with recycled compost was also found to aid drying, but such was not the case with covering windrows.

12. Finished compost monitoring at the Los Angeles facility includes total and total volatile solids, nitrogen, cadmium, lead, PCBs, and biological agents. Demonstration study monitoring at the Denver facility was varied. Levels of fecal coliform, fecal streptococci, and *A. fumigatus aerospora* greater than 10^2, 10^4, and 10^4 per gram were typical, and *Salmonella spp.* were also detected. Facility personnel report that the principal goal of the demonstration project was to study composting and produce a dried product rather than to produce a finished compost meeting specific quality requirements. Los Angeles contracts with a private firm (the Kellogg Supply Company) for sale and distribution of its product.

13. As noted in item 3, the potential for odor complaints limits the sludge quantity (500 wtpd) which can be composted at the Los Angeles facility during summer periods. In addition, personnel at this location have performed studies which indicate that 83 percent of the odor emissions from the windrow operation are the result of ambient surface emissions and 17 percent are the result of windrow turnings. It was also found that the intensity of ambient surface emissions decrease as the composting period progresses and that emissions are greatest immediately after windrow turning. Teardown of windrows during unfavorable weather conditions, e.g., inversions, creates occasional problems. At the Denver facility, odor at the site boundary infrequently exceeded a threshold value of 15 dilutions.

14. At the Los Angeles facility, dust is generated from truck unloading, mixing, and windrow turning operations and sometimes directly from windrow surfaces. *A. fumigatus aerospora* are not a problem at this location. Compost stockpiling, windrow turning, and aeration blowers with negative aeration were identified as on-site sources of *A. fumigatus aerospora* in a study commissioned by the Denver facility. Sidestreams are collected and discharged to local sanitary sewers at both windrow facilities.

REFERENCES

1. Caballero, R. "Experience at a Windrow Composting Facility: Los Angeles County Site. In Sludge Composting and Improved Incinerator Performance, EPA Technology Transfer. July 1984.

2. Hay, J.C., et al. "Disinfection of Sewage Sludge by Windrow Composting." National Science Foundation Workshop on Disinfection, Coral Gables, Florida. May 7-9, 1984.

3. Iacoboni, M.D., et al. "Windrow and Static Pile Composting of Municipal Sewage Sludges." EPA, Municipal Environmental Research Laboratory, Cincinnati, Ohio. May 1982.

4. Garrison, W.E. "Composting and Sludge Disposal Operations at the Joint Water Pollution Control Plant." County Sanitation Districts of Los Angeles County, Whittier, California. June 23, 1983.

5. LeBrun, T.J., et al. "Overview of Compost Research Conducted by the Los Angeles County Sanitation Districts." National Conference on Municipal and Industrial Sludge Composting, Philadelphia, Pennsylvania. November 17-19, 1980.

6. Hay, J.C., et al. "Forced-Aerated Windrow Composting of Sewage Sludge." Virginia Water Pollution Control Association Conference, Williamsburg, Virginia. April 30-May 2, 1984.

7. Iacoboni, M. "Compost Economics in California." Biocycle, July-August 1983.

8. Iacaboni, M., et al. "Deep Windrow Composting of Dewatered Sewage Sludge." National Conference on Municipal and Industrial Sludge Composting, Philadelphia, Pennsylvania. November 17-19, 1980.

9. Horvath, R.W. "Operating and Design Criteria for Windrow Composting of Sludge." National Conference on Design of Municipal Sludge Compost Facilities, Chicago, Illinois. August 29-31, 1978.

10. Central Plant Facility Plan, Volume IV. Prepared for the Metropolitan Denver Sewage Disposal District Number 1 by Black and Veatch Engineers-Architects (October 1983).

11. Value Engineering Study. Prepared for the Metropolitan Denver Sewage Disposal District Number 1 by Culp/Wesner/Culp, Consulting Engineers (February 23-27, 1981). As reported in Reference 10.

7. Cost Evaluation

This chapter presents an analysis of capital and operation and maintenance (O&M) costs for the static pile and windrow composting technologies studied during the technology evaluation. The analysis is based on information provided by personnel at four of the five facilities investigated: the Hampton Road Sanitation District Peninsula Composting Facility (Hampton Roads facility) located in Newport News, Virginia; the Washington Suburban Sanitary Commission Montgomery County Composting Facility (Site II facility) located in Silver Spring, Maryland; the City of Columbus, Ohio, Southwesterly Composting Facility (Columbus facility) located in Franklin County, Ohio; and the Joint Water Pollution Control Plant Composting Facility of the Los Angeles County Sanitation Districts (Los Angeles facility) located in Carson, California. Costs for the Metropolitan Denver Sewage Disposal District Number One Demonstration Composting Facility (Denver facility) located in Denver, Colorado, were not analyzed because the plant was constructed and operated for demonstration study purposes.

The chapter is organized into four subsections. Capital expenditures are presented in the first subsection; operating costs and revenues from compost sales are presented in the second and third, respectively; and a postconstruction comparison of costs is presented in the fourth.

CAPITAL EXPENDITURES

The Columbus, Hampton Roads, and Site II static pile facilities were initially constructed in 1980, 1981, and 1983, respectively (Chapter 3). Since initial development, additional capital expenditures have been required at each installation to improve and/or modify operations. The Los Angeles facility (conventional windrow) has evolved gradually since composting activites were initiated in the early 1970s, as described in Chapter 4.

Actual capital costs for the three static pile facilities are summarized by year in Table 29. As noted in the table, capital costs do not include expenditures for engineering, administration and interest during construction. Additionally, land costs, if any, and capital costs for sludge transport equipment are not included.

Actual capital costs presented in Table 29 were adjusted to midyear 1985 using the following applicable midyear _Engineering News-Record_ (ENR) 20-cities Construction Cost Indices (CCI):

TABLE 29. CAPITAL COSTS FOR STATIC PILE FACILITIES

Facility	Year[a]	Capital cost, thousand dollars[b]	
		Actual	Adjusted to 1985[c]
Hampton Roads	1981	1,068[d]	1,260
	1982	143	157
	1983	1,091	1,124
	1984	30	24
	Total	-	2,565
Columbus	1980	1,268	1,648
	1982	1,486	1,635
	1983	3,230	3,327
	Total	-	6,610
Site II	1983	16,703[e]	17,204
	1984	493	404
	Total	-	17,608

[a]First year is date of initial construction.

[b]Cost of engineering, administration and interest during construction are not included.

[c]Adjusted based on ENR 20-Cities Construction Cost Indices for midyear 1980 through 1985. See text.

[d]Excludes $255,000 for purchase of sludge transport trucks.

[e]Excludes land cost and cost for on-site, mobile equipment.

Year	ENR CCI
1977	2541
1979	2980
1980	3260
1981	3562
1982	3845
1983	4114
1984	5112
1985	4220

Based on this adjustment, the total capital cost of the 50-wtpd Hampton Roads facility in 1985 dollars is about $2,600,000. Capital costs in 1985 dollars for the 200-wtpd Columbus facility and the 400-wtpd Site II facility are about $6,600,000 and $17,600,000, respectively. Features provided at each facility vary widely because of site-specific requirements, as described in Chapter 3.

Except for equipment costs, capital expenditures for the Los Angeles facility were not readily available during the technology evaluation. Because of this, a comparison to static pile capital costs cannot be made. Facility personnel have determined that equipment capital recovery is equal to $1,050 per operating day (6-day-per-week operation), which is equivalent to $327,600 per year or about $2 per wet ton for a sludge loading of 500 wtpd. At a dewatered sludge total solids content of 24 percent, this equipment capital recovery cost translates to about $9 per dry ton.

OPERATING COSTS

Annual O&M costs in 1985 dollars for the three static pile facilities and the Los Angeles conventional windrow facility are presented in Table 30. Costs shown were calculated from actual costs based on the following applicable midyear, U.S.-average Consumer Price Indices (CPI) for urban wage earners and clerical workers:

Year	CPI
1977	182.6
1979	217.7
1983	297.2
1984	306.2
1985	322.3

On-site O&M costs are broken down into several cost categories: labor, bulking agent and aeration piping for static pile systems, external amendments for the conventional windrow system, utilities, fuel, equipment maintenance, and laboratory or other expenses. Costs for administration and sludge transport, where available or applicable, are shown as separate line items.

Annual O&M costs in 1985 dollars for on-site operations and administration at the Hampton Roads, Columbus, and Site II static pile facilities are approximately $467,000, $1,048,000, and $3,483,000, respectively. Note that fuel costs for the Columbus facility were not available, and that the cost of

TABLE 30. ANNUAL OPERATION AND MAINTENANCE COSTS

Cost category	Operation and maintenance costs adjusted to 1985, thousand dollars[a]			
	Hampton Roads	Columbus	Site II	Los Angeles
On-site operation and maintenance				
Labor, including fringe benefits	195.0	497.5	828.5	624.0
Bulking agent and aeration piping	179.0	210.0	803.0[b]	-
External amendments				0.0[c]
Utilities	3.0	35.5	191.0	-
Fuel	22.5	-[d]	-[e]	148.0
Equipment maintenance	49.5	139.0	-[e]	156.0
Laboratory and other expenses	2.5[f]	83.0[f]	109.0[g]	60.0
Subtotal, on-site operation and maintenance	451.5	965.0	1,931.5	988.0
Administration	15.5[h]	82.5	1,551.0[i]	-[j]
Subtotal, excluding sludge transport	467.0	1,047.5	3,482.5	-
Long-haul sludge transport	107.0[k]	-[l]	275.0[l]	-[m]
Total operation and maintenance	574.0	-	3,757.5	-

[a]Rounded to nearest 500 dollars.

[b]Most of this cost is for bulking agent (new wood chips, which are typically 0 to 40 percent of total bulking agent volume requirements). Other included costs are for asphalt repair, janitorial services, protective clothing, and related expenses.

[c]Amendments are used in the windrow operation but they are obtained at no cost.

[d]Not determined during technology evaluation.

[e]Fuel and maintenance cost is included in cost for administration.

[f]Analytical work is performed at the associated treatment plant. Cost is not included.

[g]Includes rental of a standby generator ($67,000).

[h]Labor only, including a part-time agronomist for compost marketing.

[i]Includes cost of fuel and equipment, per footnote "e."

[j]Management and administration is provided by personnel at the Joint Water Pollution Control Plant.

[k]Labor and fuel only. Maintenance cost was not determined during the technology evaluation.

[l]Provided by private contractor. Cost for the Columbus facility was not determined during the technology evaluation.

[m]Compost facility is located adjacent to the treatment plant so there is no long-haul sludge transport cost.

a standby generator is included at the Site II facility. Also, because analytical services at the Hampton Roads and Columbus facilities are performed at the adjacent treatment plant, costs for these services are not included.

The information presented in Table 30 shows that costs for on-site labor, including fringe benefits, comprise about 24 to 50 percent of the annual static pile O&M costs at these facilities, excluding sludge transport costs. Costs for bulking agent and aeration piping vary from about 20 to 40 percent of the annual O&M costs, excluding sludge transport. As noted previously, bulking agent (wood chip) requirements are greater than initially projected. Each facility attempts to maximize recovery and/or reuse of wood chips to keep new wood chip purchase to a minimum. Combined costs for on-site labor, new wood chip purchase, and aeration piping account for 50 to 80 percent of the annual O&M costs, excluding sludge transport, at the static pile facilities investigated. Administrative costs vary because administrative charges, e.g., fuel costs, are allocated in different ways at each facility.

The annual on-site O&M cost at the Los Angeles facility in 1985 dollar is $988,000 (Table 30). Utilities cost and administrative costs were not readily available during the technology evaluation and therefore are not included in the table. The Los Angeles facility employs external amendments as well as recycled compost for conventional windrow composting; however, there is no purchase cost for these amendments. Labor cost, at $624,000, is the major O&M cost item (63 percent), with fuel and equipment maintenance being 15 and 16 percent, respectively.

Representative annual quantities of dewatered sludge currently processed by each facility are shown below:

Facility	Sludge total solids, percent	Annual sludge quantity	
		Wet tons	Dry tons
Hampton Roads	17	14,500	2,465
Columbus	17	42,097	7,157
Site II	17	97,000	16,990
Los Angeles	23	154,752	37,140

Based on the sludge quantities presented above, unit O&M costs, excluding sludge transport, for the Hampton Roads, Columbus, and Site II facilities are about $32, $25, and $36 per wet ton in 1985 dollars, respectively. These values translate to $189, $146, and $211 per dry ton, respectively. Note (Table 30) that fuel cost at the Columbus facility was not available and therefore is not included in the unit costs presented.

Annual sludge transport costs were available for the Hampton Roads and Site II facilities (Table 30). When these costs are included, unit O&M costs in 1985 dollars increase to about $40 per wet ton ($233 per dry ton) at the Hampton Roads facility and to about $39 per wet ton ($228 per dry ton) at the Site II facility.

Unit costs for the Los Angeles conventional windrow facility are not directly comparable to those for the static pile facilities because of different facility capacities, and because O&M costs for all cost categories presented in Table 30 are not available at each of the facilities. Also, although the Los Angeles facility employs external bulking agents, there is no purchase cost for these materials. Based on the costs presented in Table 30 and the annual quantities of sludge presented above, the unit O&M cost in 1985 dollars at the Los Angeles facility is about $7 per wet ton, or $27 per dry ton.

COMPOST SALES REVENUE

Annual O&M and unit O&M costs presented in the previous section do not reflect revenues generated from the sale of finished compost at each facility. Table 31 summarizes annual revenues from the sale of finished compost in 1985 dollars, based on actual figures available from the facilities. Actual revenues have been adjusted to 1985 dollars, where necessary, using CPI-values presented previously.

Revenue from the sale of finished compost is equivalent to about $2.50, $0.60, and $1.40 per wet ton at the Hampton Roads, Columbus, and Site II static pile facilities, respectively. Revenue at the Los Angeles windrow facility is equivalent to about $1.00 per wet ton.

When revenue from the sale of finished compost is taken into account, O&M cost at the Hampton Roads facility, excluding sludge transport, is reduced from $32 per wet ton to about $29 per wet ton in 1985 dollars. The quantity of compost sold, 5,600 cu yd, represents about 60 to 70 percent of annual finished compost production.

Net O&M costs in 1985 dollars after compost sale at the Columbus and Site II facilities are about $24 and $34 per wet ton, excluding sludge transport. At Site II, the annual quantity of compost currently sold is about 33,000 cu yd or about 34 percent of annual production. Corresponding figures for Columbus are not available. All three static pile facilities have programs to increase compost sales in the future.

The sale of compost at the Los Angeles facility reduces annual O&M cost by about $1 per wet ton, or from $7 to $6 per wet ton. As noted in Chapter 6, compost is sold to Kellogg Supply Company on a contract basis.

POSTCONSTRUCTION COST COMPARISON

A postconstruction comparison of current facility costs and those estimated during facility planning and design are presented in Table 32. Only static pile costs are presented because the Los Angeles facility evolved gradually since the early 1970s and was not the product of a specific planning and design step (Chapter 4). Costs in Table 32 have been adjusted to 1985 dollars based on the ENR CCIs and CPIs previously presented in this chapter. Sludge dewatering and transport costs are not included in the table.

TABLE 31. ANNUAL REVENUE FROM COMPOST SALE

Facility	Actual revenue		Adjusted revenue, 1985 dollars[a]	Compost quantity sold, cu yd
	Year	Dollars		
Hampton Roads	1984	35,124	36,972	5,600
Columbus	1983	23,200[b]	25,160	-
Site II	1984	129,000[c]	135,783	33,000[c]
Los Angeles	1985[d]	148,500[d]	148,500	-

[a]Based on CPIs presented in a previous section of Chapter 7.

[b]Actual revenue presented is projected from a value of $5,800 for the last 3 months of 1983 when the marketing program was initiated.

[c]Estimated from actual values for the 20-month period from April 1983 to November 1984.

[d]Estimated from information for the last quarter of 1984 and first quarter of 1985.

TABLE 32. POSTCONSTRUCTION COST COMPARISON OF STATIC PILE FACILITIES

Item	Costs adjusted to 1985, thousand dollars[a]		
	Hampton Roads	Columbus	Site II
Facility Planning and Design			
Year	1979	1980	1977
Sludge loading, wtpd	80	200	600
Estimated costs			
Capital[b]	3,950	2,201[c]	16,990[d]
Operation and maintenance[e]	544.0	-[f]	2,524.0
Current Facility			
Design loading, wtpd[g]	50	200	400
Current costs			
Capital, including modifications[h]	2,565	6,610[c]	17,608
Operation and maintenance[i,j]	467.0	1,047.5	3,482.5
Annualized cost per wet ton[k]			
Design estimates			
Capital	16	4	9
Operation and maintenance	19	-[f]	12
Total, design estimate	35	-	21
Current operation			
Capital[l]	17	11	15
Operation and maintenance[m]	32	25	36
Total, current operation	49	36	51
Sales revenue[m]	3	1	2
Net, current operation	46	35	49

[a]Based on cost adjustment indices previously presented in this chapter.
[b]Includes estimates for engineering and administration.
[c]Land cost is not included because site is on city property.
[d]Excludes cost for on-site mobile equipment. Land cost of $6,178,000 is included.
[e]Annual cost excludes sludge transport and amortized capital costs.
[f]Not estimated during facility planning.
[g]Design loadings as actually constructed.
[h]From Table 29.
[i]From Table 30. Cost excludes sludge transport.
[j]Excludes revenue from compost sale.
[k]Rounded up to nearest dollar. Annualized capital costs are based on a 20 year amortization period and 10 percent interest.
[l]Based on design loadings for the current facility and 365-day-per-year operation.
[m]Based on representative annual sludge quantities of 14,500, 42,097, and 97,000 wet tons for the Hampton Roads, Columbus and Site II facilities.

The Hampton Roads and Site II facilities were originally planned as 80- and 600-wtpd facilities, respectively (Chapter 3). Estimated capital costs in 1985 dollars for these sludge loadings were $3,950,000 and $16,990,000, respectively, and corresponding estimated annual O&M costs were $544,000 and $2,524,000. Note that capital costs include estimates for land and engineering and administration. O&M costs exclude costs for amortized capital.

The current Hampton Roads facility was ultimately constructed to compost sludge at an average loading of 50 wtpd, whereas the current Site II facility was constructed for an average loading of 400 wtpd. Actual capital costs for these facilities, including modifications made since initial construction, are $2,565,000 and $17,608,000, respectively, expressed in 1985 dollars. Corresponding annual O&M costs, excluding amortized capital, are $467,000 and $3,482,500.

The Columbus facility was designed and constructed to compost sludge at an average loading of 200 wtpd. Estimated capital cost was $2,201,00, but annual O&M cost was not projected. Actual capital cost, including modifications, is $6,610,000; and current annual O&M cost, excluding amortized capital, is $1,047,500, expressed in 1985 dollars.

Annualized costs per wet ton (unit costs) were calculated for design conditions and current operations and are presented in Table 32. Annualized capital costs are based on an amortization period of 20 years and an interest rate of 10 percent.

Unit costs for design conditions are based on design sludge loadings of 80, 200, and 600 wtpd for the Hampton Roads, Columbus, and Site II facilities, respectively. Corresponding annual sludge quantities of 29,200, 73,000, and 219,000 wet tons are assumed, based on 365-day-per-year operation. Unit capital costs for current operations are based on design loadings for the current facility, again assuming 365-day-per-year operation; however, current unit O&M costs are based on representative annual sludge quantities of 14,500, 42,097, and 97,000 wet tons actually composted at the Hampton Roads, Columbus, and Site II facilities, respectively. On a calendar-day basis, these annual quantities translate to corresponding loadings of 40, 115, and 265 wtpd. Actual annual loadings are thus about 80, 58, and 66 percent of the design loadings for the respective facilities.

The postconstruction comparison shows that current unit costs at the static pile facilities are considerably higher than those estimated during facility planning and design based on information obtained during the technology evaluation. Current total unit cost varies from $36 to $51 per wet ton, and unit capital and O&M costs vary from $11 to $17 per wet ton and $25 to $36 per wet ton, respectively. Corresponding design estimates were $21 to $35 per wet ton, $4 to $16 per wet ton, and $12 to $19 per wet ton. Revenue generated from the sale of finished compost reduces the cost of current operations by $1 to $3 per wet ton.

8. Comparison of Technologies

This chapter presents a summary of key facility features and a comparison of operating and performance features of static pile and windrow technologies based on the five facilities investigated during the technology evaluation: the Hampton Road Sanitation District Peninsula Composting Facility (Hampton Roads facility) located in Newport News, Virginia; the Washington Suburban Sanitary Commission Montgomery County Composting Facility (Site II facility) located in Silver Spring, Maryland; the City of Columbus, Ohio, Southwesterly Composting Facility (Columbus facility) located in Franklin County, Ohio; the Joint Water Pollution Control Plant Composting Facility of the Los Angeles County Sanitation Districts (Los Angeles facility) located in Carson, California; and the Metropolitan Denver Sewage Disposal District Number One Demonstration Composting Facility (Denver facility) located in Denver, Colorado. The Hampton Roads, Site II, and Columbus facilities use the static pile process to compost anaerobically digested, limed raw, and unlimed raw sludge, respectively. The Los Angeles and Denver facilities use conventional and aerated windrow processes, respectively, to compost anaerobically digested sludge. Key design and operating considerations for static pile and windrow composting technologies are identified at the end of the chapter, based on the results of the technology evaluation.

FACILITY FEATURES

Key features of each facility, based on current operation, are summarized in Table 33. All facilities were constructed between 1980 and 1983, except the Los Angeles facility which has evolved gradually since the early 1970s. Sludges composted are from activated sludge wastewater treatment plants located either adjacent to or off site from the composting facility. Sludge dewatering equipment varies, and polymers are used at all except the Site II facility. Dewatered sludge storage is used at Site II and at Los Angeles to manage loadings to the composting facility.

Sludge Transport

Truck transport of dewatered sludge is used at all five facilities. Haul distance varies, as does the type and capacity of transport vehicle. At Los Angeles, semitrailers are used to haul dewatered sludge and amendment to the active windrow composting area, whereas sludge is hauled separately at the other locations. The Site II and Columbus facilities contract for sludge transport. Dewatered sludge deliveries are coordinated with composting operations at all sites.

TABLE 33. KEY FACILITY FEATURES FOR COMPOSTING TECHNOLOGIES INVESTIGATED

Item	Composting facility				
	Hampton Roads	Site II	Columbus	Los Angeles	Denver
General characteristics					
Composting process	Extended static pile	Extended static pile	Extended static pile	Conventional windrow	Aerated and conventional windrow
Date constructed	1981	1983	1980	-[a]	1982
Location	Virginia	Maryland	Ohio	Southern California	Colorado
Operating status	Fully operational	Fully operational	Fully operational	Fully operational	Demonstration study[b]
Treatment plant location	Off-site	Off-site	Off-site	Adjacent	Adjacent
Treatment plant characteristics					
Number contributing sludge	3	1	1	1	1[c]
Type	Activated sludge	Activated sludge	Activated sludge	Activated sludge	Activated sludge
Sludge type	Anaerobically digested	Limed, raw	Unlimed, raw	Anaerobically digested	Anaerobically digested
Dewatering equipment	Belt filters	Vacuum filters	Centrifuges	Centrifuges	Centrifuges
Chemical addition	Polymers	Lime, ferric chloride	Polymers	Polymers	Polymers
Sludge storage	No	Yes	No	Yes	No
Sludge transport					
Distance, miles	6 to 35	32	10	Adjacent to site	Adjacent to site
Equipment					
Type	Open top, semi-trailer trucks	Enclosed, air tight trucks	Open top, semi-trailer trucks	Open top, semi-trailer trucks	Open top, semi-trailer trucks
Capacity	23 cu yd	20 tons	25 tons	42 cu yd	-[d]
Number	2	-[d]	-[d]	3[e]	-[d]
Delivery coordinated	Yes	Yes	Yes	Yes	NA[f]
Responsibility	Owner	Private contractor	Private contractor	Owner	NA
Operational area[g]					
Total, ac	6.2	40	37	37	NA
Unit, ac/wtpd	0.12	0.10	0.18	0.08	NA
Key site features[h]					
Mixing area	P,C	P,C	P,C	-[i]	-[i]
Active composting area	P,O	P,C	P,O	P,O	P,C and O
Drying area	P,O	P,C	-[j]	-[i]	-[i]
Screening area	P,C	P,C	-[j]	NA	NA
Curing area	U,O[k]	P,O[l]	P,O[k]	-[d]	-[d]

[a]Facility has evolved gradually from air drying operation used in 1972. See Chapter 4.
[b]Facility was designed to study the effectiveness of windrow composting at Denver. See Chapter 4.
[c]Centralized sludge treatment operations for more than one treatment plant.
[d]Not determined during technology evaluation.
[e]The same trucks are used to haul sludge and amendment mix to active composting area. See Chapter 6.
[f]NA = Not applicable for this facility.
[g]Operational area is defined as site area required for current operations, including access roads, buffer zones, support facilities, and runoff control ponds. Land available on site that is not used for active composting operations, e.g., land available for expansion, is excluded.
[h]P = paved. C = covered, partially enclosed, or fully enclosed. U = unpaved. O = open, not covered.
[i]Included in active composting area.
[j]Drying and screening problems have led to facility modifications not included in the technology evaluation. See Chapter 5.
[k]Unscreened compost is stored and cured. Finished compost is stored separately.
[l]Screened compost is cured and then stored as finished compost.
[m]Not used for sludge transport.
[n]Provided by personnel at the wastewater treatment plant.
[o]Includes operating supervision.
[p]Includes two heavy auto mechanics.
[q]Facility contracts with Kellogg Supply Company for sale and ultimate distribution. Compost products are marketed to the general public in retail stores in three states. Direct advertising is not used.
[r]Plus delivery cost of $3 to $9 per cu yd.

TABLE 33 (continued)

TABLE 33. KEY FACILITY FEATURES FOR COMPOSTING TECHNOLOGIES INVESTIGATED (CONTINUED)

Item	Composting facility				
	Hampton Roads	Site II	Columbus	Los Angeles	Denver
Storage areas					
Bulking agent, amendment	P,O	P,O	P,O	U,O	P, C and O
Unscreened compost	U,O[k]	NA	P,O[k]	NA	NA
Finished compost	P,O[k]	P,O[l]	P,C[k]	U,O	P,C and O
On-site equipment, number					
Front-end loader	5	12	8	2	1
Mobile composter	0	4	2	4	1
Brown Bear	0	0	0	0	1
Manure speader	1	0	0	0	0
Tractor and rototiller	1	0	1	0	0
Mobile screen	2	0	1	0	0
Fixed screen	0	3	1	0	0
Semitrailer truck	0	0	2[m]	3[e]	0
Dump truck	0	4	0	0	0
Trailer tipper	0	0	2	0	0
Water truck	0	0	1	1	0
Manpower utilization					
Total number	8.5	42	19	11	NA
Personnel categories					
Management/administration	-[n]	5[o]	2[o]	-[n]	NA
Operating supervision	1	-	-	1	NA
Operating staff	7	32	16[p]	10	NA
Laboratory services	-[n]	5	1	-[n]	NA
Marketing	1/2 time	-	-	-	NA
Compost distribution					
Marketing program	Yes	Yes	Yes	Yes	NA
Program management	Owner	Maryland Environmental Services	Owner	Kellogg Supply Company[q]	NA
Trade name	-	ComPRO	COM-TIL	Nitrohumus, Topper, Amend, Gromulch	NA
Delivery to user	No	Yes	No	-[q]	NA
Bulk cost, dollars/cu yd.	<6 to 7.50	3 to 4[r]	9	NA	NA
User categories					
Public	Yes	Yes	Yes	Yes	NA
Commercial	Yes	Yes	Yes	Yes	NA
Government	Yes	Yes	Yes	-[d]	NA
Retail sales (bagged)	Yes	Yes	No	Yes	NA

[a]Facility has evolved gradually from air drying operation used in 1972. See Chapter 4.
[b]Facility was designed to study the effectiveness of windrow composting at Denver. See Chapter 4.
[c]Centralized sludge treatment operations for more than one treatment plant.
[d]Not determined during technology evaluation.
[e]The same trucks are used to haul sludge and amendment mix to active composting area. See Chapter 6.
[f]NA = Not applicable for this facility.
[g]Operational area is defined as site are required for current operations, including access roads, buffer zones, support facilities, and runoff control ponds. Land available on site that is not used for active composting operations, e.g., land available for expansion, is excluded.
[h]P = paved. C = covered, partially enclosed, or fully enclosed. U = unpaved. O = open, not covered.
[i]Included in active composting area.
[j]Drying and screening problems have led to facility modifications not included in the technology evaluation. See Chapter 5.
[k]Unscreened compost is stored and cured. Finished compost is stored separately.
[l]Screened compost is cured and then stored as finished compost.
[m]Not used for sludge transport.
[n]Provided by personnel at the wastewater treatment plant.
[o]Includes operating supervision.
[p]Includes two heavy auto mechanics.
[q]Facility contracts with Kellogg Supply Company for sale and ultimate distribution. Compost products are marketed to the general public in retail stores in three states. Direct advertising is not used.
[r]Plus delivery cost of $3 to $9 per cu yd.

Operational Areas

Unit operational areas for the composting technologies studied vary from 0.08 to 0.18 ac/wtpd. These values are based on design sludge loadings of 50, 200, and 400 wtpd for the Hampton Roads, Columbus, and Site II facilities, respectively, and a maximum summer loading of 500 wtpd for the Los Angeles facility. Operational area is defined as site area actually required for current composting operations, including access roads, buffer zones, support facilities, and storage ponds for runoff control, if required. Other site area, e.g., land available for expansion, is not included.

Many factors affect the unit operational area applicable to the static pile and windrow technologies investigated. For example, the Columbus static pile facility, which has a unit operational area of 0.18 ac/wtpd, uses a large (10 ac) area for unscreened compost storage because screening operations are affected by weather and the market for distribution of finished compost is still being developed (Chapter 3). This large area is the main factor contributing to a unit operational area which is considerably higher than the other facilities investigated. The Los Angeles windrow facility uses windrows which are larger than those typically employed for conventional windrowing (Chapter 6), and this may contribute to a unit operational area that is lower than the other facilities.

Air drying by spreading and rototilling at the Hampton Roads facility requires about 1,500 sq ft/wtpd, whereas intensive prescreen drying at the Site II facility utilizes only 100 sq ft/wtpd. Drying in place is used for windrow composting and, thus, is included as part of the active composting area for this technology. Runoff containment ponds at the Site II and Columbus facilities require 1,100 and 200 sq ft/wtpd, respectively, whereas no ponds are used at the Hampton Roads facility.

If the 10-ac, unscreened compost storage area at the Columbus facility is excluded, an adjusted unit operational area of 0.14 ac/wtpd is obtained. Thus, given the diversity of size, operation, and other factors at the facilities investigated, a reasonable range of unit operational areas for the static pile and windrow technologies studied is 0.08 to 0.14 ac/wtpd.

Paving and Covering

Paved areas are used for most operations at both the static pile and windrow facilities investigated. Paving was added to the main active windrow composting area at the Los Angeles facility after lime-stabilized dirt proved unsatisfactory during wet-weather operation (Chapter 6), and similar post-construction improvements to mixing, drying, screening, curing, and/or materials storage areas were implemented in response to wet-weather problems at the Hampton Roads and Columbus locations (Chapter 5). The Site II static pile facility and the windrow demonstration study facility at Denver were constructed with the paved site features indicated in Table 33.

Covered operations are less common than paved operations at the static pile and windrow facilities investigated. Although all composting activities at the Los Angeles facility are uncovered, wet weather impacts operation and leads to winter productivities which are lower than those during dry, summer weather. Mixing, active composting, drying, and screening areas at the Site II static pile facility were covered as part of facility design, and postconstruction covered mixing has been provided at Hampton Roads and Columbus because of wet-weather problems. Postconstruction covered screening has also been provided at Hampton Roads. Covered and uncovered windrow composting were investigated during the demonstration study at the Denver facility, and an independent analysis of study data during the technology evaluation indicated that mean moisture reductions were similar for both covered and uncovered windrows. This suggests that under the conditions studied at Denver, covering the windrows did not enhance drying. Additionally, it was observed that in many cases, maximum temperatures for uncovered windrows were greater than 55 degrees C for a longer duration than for covered windrows. The reason for this was not evident from the work done during the technology evaluation.

Equipment

Front-end loaders are the only on-site equipment items common to all facilities. Both static pile and windrow technologies use mobile composters; however, only two of the three static pile facilities investigated are provided with this type of equipment (for mixing). The mobile composters at the Columbus facility are not used currently in daily operation because of uneven pavement in the mxixing area (Chapter 5). The number of front-end loaders and mobile composters varies from site to site, as do capacities and applications. Note that screening equipment is not required for process operations at the windrow facilities, and that other on-site equipment at the static pile and windrow facilities studied are site-specific.

Manpower Utilization

Manpower utilization reported by facility personnel varies at each composting site. Total number of personnel varies from 8.5 at the Hampton Roads facility to 42 at the Site II facility. Personnel utilization at Denver is not reported because this facility was constructed for demonstration study purposes. Total personnel utilization is affected by the number of management, administrative, and supervisory personnel at each location, as well as those provided for laboratory and marketing services. The number of operating personnel exclusive of these five personnel categories varies from 7 to 32.

Ten operating personnel are used at the Los Angeles windrow facility, which treats sludge at a loading of 400 to 500 wtpd, compared to the Site II static pile facility which uses 32 operating personnel and was designed for a loading of 400 wtpd. The Columbus static pile facility, which was designed at an annual average sludge loading of 200 wtpd, uses 16 operating personnel. Based on this comparison, the static pile technology is more labor intensive than the conventional windrow technology. Relative skill levels of operating personnel were not assessed during the technology evaluation.

Compost Distribution

All fully operational facilities (excludes Denver) have marketing programs for compost sale. The Hampton Roads and Columbus facilities market their product directly, whereas independent marketing entities are used at the Site II and Los Angeles facilities. Delivery services are available only at the Site II facility.

Bulk cost varies from $3 to $9/cu yd at the static pile facilities, depending on quantity sold. An additional cost of $3 to $9/cu yd is applied for delivery at the Site II facility. The Kellogg Supply Company markets compost for retail sale in three states and also advertises to the nursery trade. Direct advertising to the general public is not used. Typical compost users at the three static pile facilities studied include the general public, commercial establishments such as nurseries, and government agencies. Bagged product is sold at all but the Columbus facility.

OPERATING FEATURES

Operations assessments for the static pile and windrow facilities investigated during the technology evaluation are presented in Chapters 5 and 6, respectively. Each chapter includes a postconstruction comparison of the respective technology. This chapter identifies and compares key operating features common to both technologies.

Flexibility

Flexibility to respond to variable sludge loadings, weather conditions, and day-to-day problems is an important operating feature of both static pile and windrow technologies. Representative current sludge loadings for each facility are summarized in Table 34, and from this comparison the importance of establishing sludge loading design and operating criteria for the composting technologies studied is apparent. Variations in dewatered sludge total solids concentration, peak-to-average-day ratio, and number of operating days per week affect day-to-day loadings that must be processed, and these factors need to be adequately addressed during facility design to ensure that process, manpower, equipment, materials handling, and budgetary considerations for effective composting are met. Failure to do so may require substantial postconstruction facility modifications and can lead to odor generation, nonperformance, and higher-than-expected operating cost. Timing sludge deliveries, coordinating treatment plant operations such as dewatering with composting constraints, and providing dewatered sludge storage are examples of ways in which sludge loading variability is currently managed at the facilities studied. The need to establish realistic dewatered sludge total solids concentrations for facility design is particularly important to minimize postconstruction operating problems and facility modifications.

Moisture Control

Moisture control, combined with effective mixing, during day-to-day operations was found to be the single most important factor for effective static pile or windrow composting. Whether moisture is present in incoming

TABLE 34. SLUDGE LOADING COMPARISON

Item	Static pile facilities			Windrow facilities	
	Hampton Roads	Site II	Columbus	Los Angeles	Denver
Solids concentration, percent					
Total, average	17	17	17	23	16
Total, minimum	14	15	13	22	-[a]
Volatile, average	55	43	74	50	60
Loading per calendar day					
Wet tons, average	54	257	170	425	NA[b]
Dry tons, average	9	44	29	98	NA
Peak-to-average	1.6 to 1.8	1.4 to 1.6	1.9	1.2	NA
Operating days per week	6	5	7	6	NA
Loading per operating day					
Wet tons, average	63	360	170	495	NA
Dry tons, average	11	62	29	114	NA
Wet tons, peak	113[c]	575[c]	325	580	NA
Dry tons, peak	19[c]	100[c]	55	133	NA
Loadings, percent of design					
Wet tons, average	126	90	85	NA	NA
Dry tons, average	110	70	73	NA	NA

[a]Not determined during technology evaluation.
[b]NA = Not applicable for this facility.
[c]Based on highest peak-to-average day ratio.

dewatered sludge, the bulking agent, or amendment; whether it is added by inclement weather; or whether it is generated as a by-product of the composting process, moisture must be controlled for effective stabilization, pathogen inactivation, odor control, and finished compost quality control.

As noted in the previous section of this chapter, paved and, in some instances, covered operating areas can aid moisture control, and therefore the need for such site features needs to be carefully assessed during design. An initial mix moisture content of 60 percent or less (total solids $\geq$40 percent), coupled with mix homogeneity, is a key criterion for effective composting regardless of the composting technology employed.

Drying in place during windrow composting, or as a separate step during static pile composting, generally must meet a performance criterion of 50 to 60 percent total solids based on operations at the five facilities studied. Effective screening at static pile facilities requires a minimum unscreened compost total solids content of 50 to 55 percent, while the goal for final total solids of compost which is recycled as an amendment during windrow composting may be as high as 60 percent. (Screening for amendment recovery is not employed during windrow composting.) Induced aeration during windrow composting (aerated windrow) can improve drying by about 2 to 3 percent based on demonstration study results from the Denver facility.

Equipment Utilization

Equipment utilization for the static pile and windrow technologies studied is compared in Table 35. Open-top, semitrailers are most commonly used for sludge transport, although the Site II static pile facility is an exception. Front-end loaders are used for a number of static pile and windrow operations, including (1) rough mixing at four of the five facilities; (2) pile construction and teardown; (3) windrow formation and teardown; and (4) some, or all, materials transport or transfer operations. Mobile composters are used (or are available for use) as fine mixers at two of the static pile facilities and both windrow facilities. Equipment used for other operations is site-specific. Separate drying and screening equipment is utilized only for the static pile technology.

Odor Control

Odor generation and release have periodically created problems and/or exceeded dilution thresholds at all composting facilities investigated, as described in Chapters 5 and 6. Each installation has facilities and/or operating strategies for odor control.

Technical and public relations issues associated with odor control are complex since citizen complaints may be either in direct response to odor released from a composting facility, or from lingering public opposition to facility siting rather than *a priori* evidence of such release. Such issues need to be carefully and systematically investigated where appropriate. Effective public relations programs are important in this regard.

TABLE 35. COMPARISON OF EQUIPMENT UTILIZATION

Key operations and equipment[a]	Static pile facilities			Windrow facilities	
	Hampton Roads	Site II	Columbus	Los Angeles	Denver
Sludge transport					
Horizontal-discharge, semitrailer truck open top	X	-	-	X	X
End-dump, semitrailer truck, open top	-	-	X	X	-
Completely enclosed truck	-	X	-	-	-
Mixing operations					
Rough mix					
Front-end loader	X	X	X	X	-
Brown Bear	-	-	-	-	X
Final mix					
Front-end loader	-	-	X	-	-
Mobile composter	-	X	-[b]	X	X
Manure spreader	X	-	-	-	-
Rototiller	X	-	-	-	-
Pile construction/windrow formation[c]					
Front-end loader	X	X	X	X	X
Mobile composter	-	-	-	X[d]	X[d]
Brown Bear	-	-	-	-	X
Pile/windrow teardown					
Front-end loader	X	X	X	X	X
General on-site materials transport or transfer[e]					
Front-end loader	X	X	X	X	X
Semitrailer truck, open top	-	-	X	X	-
Dump truck	-	X	-	-	-
Trailer tipper	-	-	X	-	-
Conveyor	-	-	X[f]	-	-
Drying and screening[g]					
Rototiller (drying)	X	-	-	-	-
Mobile, vibratory-deck screen	X	-	-	-	-
Mobile, rotary screen	-	-	X	-	-
Fixed, vibratory-deck screen	-	X	X	-	-

[a]Usual operation.
[b]Design specified mobile composters. They are not used because of uneven pavement in the mixing area.
[c]Pile construction at static pile facilities includes laying base material, constructing extended pile compartment, and adding cover material. Windrow formation operations at Los Angeles and Denver are described in Chapter 6.
[d]Includes aeration and mixing during windrow composting.
[e]See text for specific materials transport and transfer operations.
[f]Conveyors have been installed for some operations at this facility, but were not being used during the technology evaluation.
[g]Static pile facilities only.

Properties of Compost Materials

Key properties of compost materials at the static pile and windrow facilities studied are summarized in Table 36. Values presented are based on independent measurements obtained during the technology evaluation, as well as routine plant monitoring.

PERFORMANCE COMPARISON

In general, the operating static pile and windrow facilities (excludes the Denver facility) investigated during the technology evaluation routinely meet pathogen inactivation and volatile solids reduction requirements and produce a marketable product. Finished compost quality varies at these facilities, in part, because of varying sludge characteristics, e.g., metal content, and, in part, because of marketing requirements, e.g., the Los Angeles facility.

Long-term performance based on finished compost production at two static pile facilities and one windrow facility is summarized below:

	Finished compost production	
Facility	Cu yd compost per wet ton sludge	Dry tons compost per dry ton sludge
Hampton Roads	0.5-0.7	0.7-0.9
Site II	1	1.6-2.2
Los Angeles	0.4	0.7

Representative finished compost total solids contents for the Hampton Roads, Site II, and Los Angeles facilities are 52, 51, and 60 percent, respectively, and corresponding bulk densities are 900, 1,100 to 1,500, and 1,300 lb/cu yd. Typical dewatered sludge total solids content, as received, for the Hampton Roads and Site II facilities is 17 percent, whereas that for the Los Angeles facility is 23 percent. Los Angeles facility data are based on recycling finished compost without external amendments.

The finished compost production rates presented above represent three types of composting operations. One facility employs the extended aerated static pile process to compost anaerobically digested primary and secondary sludge (Hampton Roads facility); another composts a mixture of raw, limed primary and secondary solids utilizing the same process (Site II facility); and the third (Los Angeles facility) employs the conventional windrow technique to compost anaerobically digested primary and secondary solids.

Finished compost production for the static pile facility processing raw, limed sludge is greater than that for the static pile facility processing anaerobically digested sludge. Dewatered sludge and finished compost total solids are similar for each facility; however, the former handles about 350 to 400 wtpd while the latter processes about 50 to 60 wtpd. Finished compost bulk density for the facility processing raw, limed sludge is greater than that for the facility processing anaerobically digested sludge.

TABLE 36. KEY PROPERTIES OF COMPOST MATERIALS

Material	Solids, percent		Bulk density, lb/cu yd
	Total	Total volatile	
Dewatered sludge			
Anaerobically digested	16-17	54-60	1,500-1,600
Anaerobically digested	23	50	1,800
Raw, unlimed	17	71-74	1,900
Raw, limed	17	43	-
Wood chips			
Fresh	56-58	71-98	450-670
Screened	55	82	770
Wood chip-sludge mix			
Digested sludge	41	69	830
Raw, unlimed sludge	39	73	1,230
Unscreened wood chips			
Dry (digested sludge composting)	56	59	875
Wet (digested sludge composting)	48	68	930
Wet (raw, unlimed sludge composting)	40-44	63-66	1,090-1,120
Finished compost			
Static pile (digested sludge composting)	52	49	900
Static pile (raw, unlimed sludge composting)	52-69	46-64	1,100-1,400
Static pile (raw, limed sludge composting)	51	-	1,000-1,500
Windrow (digested sludge composting)			
Recycled compost amendment	60-65	40-45	1,300
Sawdust or rice hull amendment	50-55	-	850-875

Finished compost production for the aerated static pile and conventional windrow facilities processing anaerobically digested sludge are similar. The conventional windrow system processes about 425 wtpd of sludge, while the static pile facility treats 50 to 60 wtpd. Dewatered sludge total solids content of the conventional windrow facility, at 23 percent, is higher than that of the static pile facility (17 percent).

Although the long-term finished compost production figures presented above are useful for general comparisons, operational variables affect actual day-to-day output. This is illustrated by the independent materials balance analysis which was performed at the Columbus facility during the technology evaluation (Chapter 5). As noted in that chapter, finished (screened) compost production at this facility was only 0.2 dry tons per dry ton of sludge, based on the materials balance analysis. This translates to about 0.05 cu yd of finished compost per wet ton of sludge, based on a finished compost bulk density of about 1,100 to 1,200 lb/cu yd and total solids contents of 52 and 16 percent for finished compost and dewatered sludge, respectively. As described in Chapter 5, screening at this static pile facility, which processes raw, unlimed primary and secondary sludge, is not representative of steady state operations.

DESIGN AND OPERATING CONSIDERATIONS

Operating experience at the static pile and windrow facilities investigated during the technology evaluation identified many factors which should be considered prior to construction of a composting facility using these technologies. A check list of key factors is presented in Appendix C.

Appendix A: Inventory of Full-Scale Municipal Sludge Composting Facilities

This inventory was prepared for the U.S. Environmental Protection Agency as part of a comprehensive program to perform post construction evaluations of municipal composting facilities located in North America. Facilities of interest are those that are full scale and that compost municipal sewage sludge.

The inventory consists of a table identified as:

Table A1--Inventory of Operating Municipal Sludge Composting Facilities.

This table includes the location, process and length of operation for each facility. Information in this table was obtained from EPA regional sludge coordinators, state coordinators and the composting facilities.

TABLE A1. INVENTORY OF OPERATING MUNICIPAL SLUDGE COMPOSTING FACILITIES

EPA REGION	PLANT DESCRIPTION	COMPOSTING PROCESS USED	TYPE OF SLUDGE COMPOSTED	SLUDGE DEWATERING PROCESS	PERCENT SOLIDS	COMPOSTING CAPACITY, DRY TONS/DAY	YEARS OPERATING	NOTES
I	GREENWICH, CONNECTICUT FRANK FIUMARA 203-622-6447	AERATED PILE	ANAEROBICALLY DIGESTED	VACUUM FILTER	10 - 15	0.3 - 1.3		
I	BANGOR, MAINE RALPH NISHOU 207-947-0341	AERATED PILE	PRIMARY	VACUUM FILTER		11 1/2, 1 DAY/WEEK	8	
I	BAR HARBOR, MAINE BRIAN KENE 207-288-4028	AERATED PILE	SECONDARY	GRAVITY THICKENER	10	1.4	3	POLYMER ADDED TO THICKENER. OPERATES MAY TO NOVEMBER ONLY.
I	GARDNER, MAINE TIMOTHY LEVASSEUR 207-582-1351	AERATED PILE	SECONDARY (RBC) AND SCREENINGS	BELT FILTER PRESS	18 - 20	2.0, 3 DAYS/WEEK	1.5	99% OF COMPOSTED SLUDGE IS SECONDARY SLUDGE.
I	KITTERY, MAINE LESLEY DORR 207-439-4646	AERATED PILE	SECONDARY (ACTIVATED SLUDGE)	CENTRIFUGE		<1.0	3	
I	OLD ORCHARD BEACH, MAINE ED COREAU 207-934-5754	AERATED PILE	PRIMARY	SLUDGE PRESS	12	1.0 - 2.5	4	COMPOSTING FACILITY IS SEPARATE FROM WASTEWATER FACILITY.
I	OLD TOWN, MAINE LEO ROBICHAUD 207-827-5535	AERATED PILE	SECONDARY (RBC)	BELT FILTER PRESS	12 - 13	4.0, 1 DAY/WEEK	5	WOOD CHIPS USED AS BULKING AGENT.
I	SOUTH PORTLAND, MAINE GARY O'CONNEL 207-799-2552	AERATED PILE	SECONDARY (ACTIVATED SLUDGE)	COIL FILTER	15	3.0	START-UP	TRIED START-UP LAST SUMMER, HAD PROBLEMS IN WINTER. WILL START UP AGAIN SOON.
I	PORTLAND, MAINE KEN EVERETT 207-773-7785	AERATED PILE	SECONDARY (ACTIVATED SLUDGE)	BELT FILTER PRESS		2.0	2	OPERATES ONE MONTH PER YEAR ONLY. COMPOSTS SLUDGE FROM WESTBROOK WASTEWATER FACILITY.
I	WINTER HARBOR, MAINE ROBERT SOTIROS 207-581-7500	AERATED PILE		NO	INFORMATION	AVAILABLE		

TABLE A1 (continued)

TABLE A1. INVENTORY OF OPERATING MUNICIPAL SLUDGE COMPOSTING FACILITIES

EPA REGION	PLANT DESCRIPTION	COMPOSTING PROCESS USED	TYPE OF SLUDGE COMPOSTED	SLUDGE DEWATERING PROCESS	PERCENT SOLIDS	COMPOSTING CAPACITY, DRY TONS/DAY	YEARS OPERATING	NOTES
I	YARMOUTH, MAINE GIL ST. PIERRE 207-846-4803	WINDROW	SECONDARY (ACTIVATED SLUDGE)	DCG (GRAVITY THICKENER)	6 - 10	2.5	10	
I	SWAMPSCOTT, MASSACHUSETTS 617-592-5393		NO	INFORMATION	AVAILABLE			
I	DURHAM, NEW HAMPSHIRE JOHN JACKMAN 602-868-2274	AERATED PILE	PRIMARY AND SECONDARY (ACTIVATED SLUDGE)	COIL FILTER, WITH LIME	18 - 22	1.25 2 DAYS/WEEK	8	COMPOST 8.5 MONTHS/YEAR. STORE SLUDGE FOR 3.5 MONTHS/YEAR. NORMALLY 75% SECONDARY. WOOD CHIPS USED AS BULKING AGENT
I	LITTLETOWN, NEW HAMPSHIRE ROBERT LAGACE 603-444-5400	AERATED PILE	PRIMARY AND SECONDARY	COIL FILTER		0.1	2	4 DRY TONS/MONTH.
I	MERRIMACK, NEW HAMPSHIRE LARRY SPENCER 603-883-8196	AERATED PILE	PRIMARY AND SECONDARY (ACTIVATED SLUDGE)	BELT FILTER PRESS	13 - 25	5 - 7 5 DAYS/WEEK	2	WOOD CHIPS USED AS BULKING AGENT.
I	PLYMOUTH, NEW HAMPSHIRE JOHN WOOD 603-536-1733	AERATED PILE	PRIMARY	COIL FILTER	20 - 30	0.25	5	WOOD CHIPS USED AS BULKING AGENT IN RATIO 2:1. TWO 300 CUBIC YARD PILES ARE MADE PER YEAR.
I	WEST WARWICK, RHODE ISLAND METCALF & EDDY 401-822-8228	AERATED PILE	PRIMARY AND SECONDARY (DAFT FLOAT)	VACUUM COIL FILTER		2.5	7	SLUDGE IS TREATED WITH LIME AND FERRIC CHLORIDE.
II	CAPE MAY COUNTY, NEW JERSEY NABIL HANNA 609-465-9026	ENCLOSED REACTOR, ABV SYSTEM	PRIMARY AND SECONDARY	BELT FILTER PRESS WITH POLYMER	25 - 30	12	CONSTRUCTION, WILL START UP OCTOBER, 1984	WILL COMPOST SECONDARY SLUDGE FROM TWO EXISTING PLANTS AND PRIMARY SLUDGE FROM ONE NEW PLANT.
II	MIDDLETOWN, NEW JERSEY DENNIS BRODERCKI 201-495-1010	AERATED PILE	DIGESTED PRIMARY AND SECONDARY	BELT FILTER PRESS		6.7	2 - 3	POST CONSTRUCTION EVALUATION BY ROY F. WESTON, INC. (1983)
III	CAMBRIDGE, MARYLAND GUNTHER SPOHR 301-228-4496	AERATED PILE	PRIMARY AND SECONDARY (ACTIVATED SLUDGE)	VACUUM FILTER	10 - 20	20 - 30 3 DAYS/WEEK	4	SLUDGE IS LIME TREATED. POLE PEELINGS USED AS BULKING AGENT.

TABLE A1 (continued)

TABLE A1. INVENTORY OF OPERATING MUNICIPAL SLUDGE COMPOSTING FACILITIES

EPA REGION	PLANT DESCRIPTION	COMPOSTING PROCESS USED	TYPE OF SLUDGE COMPOSTED	SLUDGE DEWATERING PROCESS	PERCENT SOLIDS	COMPOSTING CAPACITY, DRY TONS/DAY	YEARS OPERATING	NOTES
III	MONTGOMERY COUNTY/ DICKERSON, MARYLAND SITE II CHUCK MURRAY 301-888-4883	AERATED PILE	PRIMARY AND SECONDARY	VACUUM FILTER WITH FERRIC CHLORIDE AND LIME	18	60 5 DAYS/WEEK	4	SITE II HAS BEEN IN OPERATION SINCE 1983. INTERIM SITE AT DICKERSON USED 1980 TO 1983
III	SHATARA, PENNSYLVANIA GARY KREMER 717-566-3361	AERATED PILE	PRIMARY AND SECONDARY	CENTRIFUGE WITH POLYMER		1.5	0.5	COMPOSTING IS DONE OFF SITE BY A PRIVATE COMPANY, ARMSTRONG AND SONS.
III	PHILADELPHIA, PENNSYLVANIA FRANK SENSKE 215-686-3864	AERATED PILE	ANAEROBICALLY DIGESTED PRIMARY AND SECONDARY	CENTRIFUGE AND BELT FILTER PRESS	20	125 - 180	3 -4	
III	LORTON, FAIRFAX COUNTY, VIRGINIA MR. SMITHBURGER 703-690-3047	AERATED PILE	PRIMARY, SECONDARY, AND ANAEROBICALLY DIGESTED	DEWATERED SLUDGE IS RECEIVED FROM ALEXANDRIA AND BLUE PLAINS	22 - 24	60 6 DAYS/WEEK	0.5	COMPOST FACILITY IS SEPARATE FROM WASTE WATER TREATMENT PLANT.
III	NEWPORT NEWS, HAMPTON RHODES SANITATION DISTRICT VIRGINIA TODD WILLIAMS 804-877-8356	AERATED PILE	ANAEROBICALLY DIGESTED PRIMARY AND SECONDARY (AC-TIVATED SLUDGE)	BELT FILTER PRESS WITH POLYMER	16 - 17	10	3	AIR IS DRAWN IN AND BLOWN OUT OF PILE. WOODCHIPS ARE USED AS BULKING AGENT.
III	UPPER OCCOQUAN, VIRGINIA MILLARD ROBBINS JR. 703-830-2200	COVERED AERATED WINDROW	ANAEROBICALLY DIGESTED PRIMARY, SECONDARY, AND TERTIARY	RECESSED TRAY FILTER PRESS	40	4	5.5	TERTIARY SLUDGE IS CHEMICAL SLUDGE FROM ADVANCED WASTE WATER TREATMENT.
III	BLUE PLAINS, DISTRICT OF COLUMBIA EARL SASSER 202-767-7641	AERATED PILE	LIME TREATED PRIMARY AND SECONDARY (ACTIVATED SLUDGE)	VACUUM FILTER	18 - 20	30	6	NOW AT 30DT/DAY, WAS AT 20DT. WILL BE UP TO 72 IN SEPT. 1984. EACH MUNICIPALITY SERVED BY PLANT IS RESPON-SIBLE FOR THEIR PORTION OF SLUDGE. SOME SENT TO FAIRFAX CO., SOME TO SITE II.
IV	BROWARD COUNTY, FLORIDA							INFORMATION REQUESTED BUT NOT RECEIVED.
IV	HOLLYWOOD, FLORIDA							INFORMATION REQUESTED.BUT NOT RECEIVED.
IV	MORGANTON, NORTH CAROLINA CARL HENNESSEE 704-433-0280	AERATED PILE	SECONDARY (PURE OXYGEN ACTIVATED SLUDGE)	CENTRIFUGES WITH POLYMER	12		4	

TABLE A1 (continued)

TABLE A1. INVENTORY OF OPERATING MUNICIPAL SLUDGE COMPOSTING FACILITIES

EPA REGION	PLANT DESCRIPTION	COMPOSTING PROCESS USED	TYPE OF SLUDGE COMPOSTED	SLUDGE DEWATERING PROCESS	PERCENT SOLIDS	COMPOSTING CAPACITY, DRY TONS/DAY	YEARS OPERATING	NOTES
IV	MYRTLE BEACH, NORTH CAROLINA							INFORMATION REQUESTED BUT NOT RECEIVED.
IV	DRY CREEK, NASHVILLE TENNESSEE FRED CLINARD 615-259-6425	AERATED PILE	AEROBICALLY DIGESTED	VACUUM PRESS WITH POLYMER	13	1.3	1	WOODCHIPS USED AS BULKING AGENT.
V	PLUCHER POOLE, INDIANA TED FLYNN 812-876-4875	WINDROW	PRIMARY AND SECONDARY	VACUUM FILTER		2 - 3	1	
V	PINE RIVER, MINNESOTA KEITH ARBURIDOS 218-587-2824	AERATED PILE	SECONDARY (RBC)	SLUDGE PRESS		2.0 1 DAY/WEEK	1	FACILITY NOT OPERATING WELL YET. COMPOST ONCE PER WEEK WOOD CHIPS USED AS BULKING AGENTS.
V	SOUTHWEST COMPOST FACILITY, COLUMBUS, OHIO ROBERT SMITH 614-222-7610	AERATED PILE	PRIMARY AND SECONDARY (ACTIVATED SLUDGE)	CENTRIFUGE WITH POLYMER	15 - 18	40	3	WOOD CHIPS USED AS BULKING AGENT
VI	HASKEL, EL PASO, TEXAS ROBERTO GUSTAMONTE 915-541-4000	WINDROW	ANAEROBICALLY DIGESTED PRIMARY AND TRICKLING FILTER SLUDGE. AEROBICALLY DIGESTED WASTE ACTIVATED	CENTRIFUGE		17	25	
VI	SOCORRO, EL PASO, TEXAS ROBERTO GUSTAMONTE 915-541-4000	WINDROW	ANAEROBICALLY DIGESTED PRIMARY	DRYING BEDS		8	12	SAWDUST USED AS BULKING AGENT, ADDED IN 1 TO 1 RATIO. NO CHEMICAL ADDITION.
VII	TOPEKA, KANSAS EDDIE SNETHAN 913-295-3873	WINDROW	ANAEROBICALLY DIGESTED SECONDARY	BELT FILTER PRESS		2 - 3	3	
VII	KEARNEY, NEBRASKA GEORGE FAIRFIELD 308-237-5133	WINDROW	PRIMARY	COIL FILTER		2	5	
VII	BEATRICE, NEBRASKA BRUCE BATES 402-223-2101	WINDROW	PRIMARY AND SECONDARY	BELT FILTER PRESS		6 2 DAYS/WEEK	1	

TABLE A1 (continued)

TABLE A1. INVENTORY OF OPERATING MUNICIPAL SLUDGE COMPOSTING FACILITIES

EPA REGION	PLANT DESCRIPTION	COMPOSTING PROCESS USED	TYPE OF SLUDGE COMPOSTED	SLUDGE DEWATERING PROCESS	PERCENT SOLIDS	COMPOSTING CAPACITY, DRY TONS/DAY	YEARS OPERATING	NOTES
VII	GRAND ISLAND, NEBRASKA WAYNE BENNETT 308-381-5440	WINDROW	PRIMARY	COIL FILTER		4	3	
VIII	EKO-KOMPOST, MISSOULA MONTANA JOSEPH C. HORVATH 406-721-1423	TO BE PATENTED	SEWAGE SLUDGE ANIMAL BLOOD BONE MEAL LEAVES	PROPRIETARY	20	82	7	28 TO 30% OF MATERIAL COMPOSTED IS SEWAGE SLUDGE. PRIVATELY OWNED AND OPERATED FACILITY.
VIII	WORLAND, WYOMING E.J. FANNING, WYOMING DEPT. OF ENVIRONMENTAL QUALITY 307-777-7074	WINDROW	20 YEAR ACCUMULATION OF LAGOON SLUDGE	NONE		-	-	PROCESS WILL BE COMPLETE THIS SUMMER WHEN LAGOON REHABILITATION IS COMPLETE.
VIII	DENVER METRO SEWER DISTRICT NO. 1, DENVER COLORADO BILL MARTIN 303-289-5941	AERATED WINDROW	ANAEROBICALLY DIGESTED PRIMARY AND SECONDARY (DAFT WAS)	CENTRIFUGE WITH POLYMER	16	10	1.5	DEMONSTRATION PLANT AT 10 dt/. UNDER DESIGN FOR 100 dt/DAY. TEST OPERATION SHOWS PROCESS NEEDS TO BE COVERED TO WORK. 7 DAY/WEEK OPERATION.
IX	OAKLAND, CALIFORNIA EAST BAY MUNICIPAL UTILITY DISTRICT, DENNIS DIEMER 415-465-3700	AERATED PILE	ANAEROBICALLY DIGESTED PRIMARY AND SECONDARY	CENTRIFUGE WITH POLYMER	20	6.5 - 16	1	
IX	LOS ANGELES, CALIFORNIA MIKE MOSHIRI 213-689-7411	WINDROW	PRIMARY	CENTRIFUGE WITH POLYMER	23 - 24	140 6 DAYS/WEEK	15	
IX	TILLO PRODUCTS CO., CITY OF SO. SAN FRANCISCO CALIFORNIA DAVE WESTERBEKE 415-588-8033	WINDROW	MUNICIPAL AND INDUSTRIAL	GRAVITY THICKENER WITH POLYMER		2.5	25	80% OF SLUDGE IS MUNICIPAL.
X	PORTLAND, OREGON	ENCLOSED REACTOR, TAULMAN WEISS	DIGESTED	BELT FILTER PRESS		60	TO START-UP APRIL - MAY 1984	DAN GUILL, TAULMAN WEISS, ATLANTA 404-262-3131
-	WINDSOR, ONTARIO, CANADA	AERATED PILE	RAW PRIMARY, TREATED WITH FERRIC CHLORIDE AND POLYMER	CENTRIFUGE	30 - 40	25 - 30 3 DAYS/WEEK	8	INCLUDES ENCLOSED POST-DRYING STEP. WOOD CHIPS AND SHREDDED RUBBER TIRES USED AS BULKING AGENT.

Appendix B: Ambient Temperature Effect on Static Pile Temperature Development

A study was conducted as part of the municipal sludge technology evaluation to assess the impact of ambient temperature on static pile temperature. Routine monitoring data from the Hampton Roads Sanitation District Peninsula Composting Facility (Hampton Roads facility) located in Newport News, Virginia, was used in the assessment.

Monitoring information reported on the Hampton Roads facility extended aerated static pile data sheets for the months of January through March and June through August 1983 were evaluated. These data sheets record initial pile construction information, including sludge type, bulking agent type, percent total and volatile solids, and the volume of each type of material. Daily monitoring includes pile temperature from six locations within the pile; percent oxygen from three locations; blower cycle, pressure, and damper settings; and the ambient temperature and precipitation. During the two 3-month periods studied, 119 piles were constructed and monitored.

Static pile temperature data were evaluated by determining the number of days that were required in order to obtain a temperature reading of 55 degrees C or greater at the six probes for 3 consecutive days. The values for each pile during each month were averaged and the arithmetic mean, standard deviation, and variance for each month were determined. The results are plotted on Figure B-1. The overall mean, deviation, and variance were also calculated for the 6-month period. The overall range was 3 to 16 days, the mean was 6.134 days with a standard deviation of 2.289 and a variance of 5.242.

Ambient temperatures that were recorded for each day of pile construction were al so analyzed to determine monthly mean ambient temperatures. Monthly mean ambient temperatues are plotted on Figure B-1 so that a comparison to the internal pile temperature data can be made. The comparison indicates that there is a direct inverse coreelation of the ambient temperature to the length of time necessary to reach 55 degrees C on 3 consecutive days.

A frequency distribution (histogram) showing the number of days required to reach 55 degrees C versus number of piles is shown on Figure B-2. Of 119 piles, 49 piles required from 5 to 7 days to reach 55 degrees C performance criterion. Twenty-seven piles required 3 to 5 days to meet the standard, and 28 piles required 7 to 9 days to meet the standard.

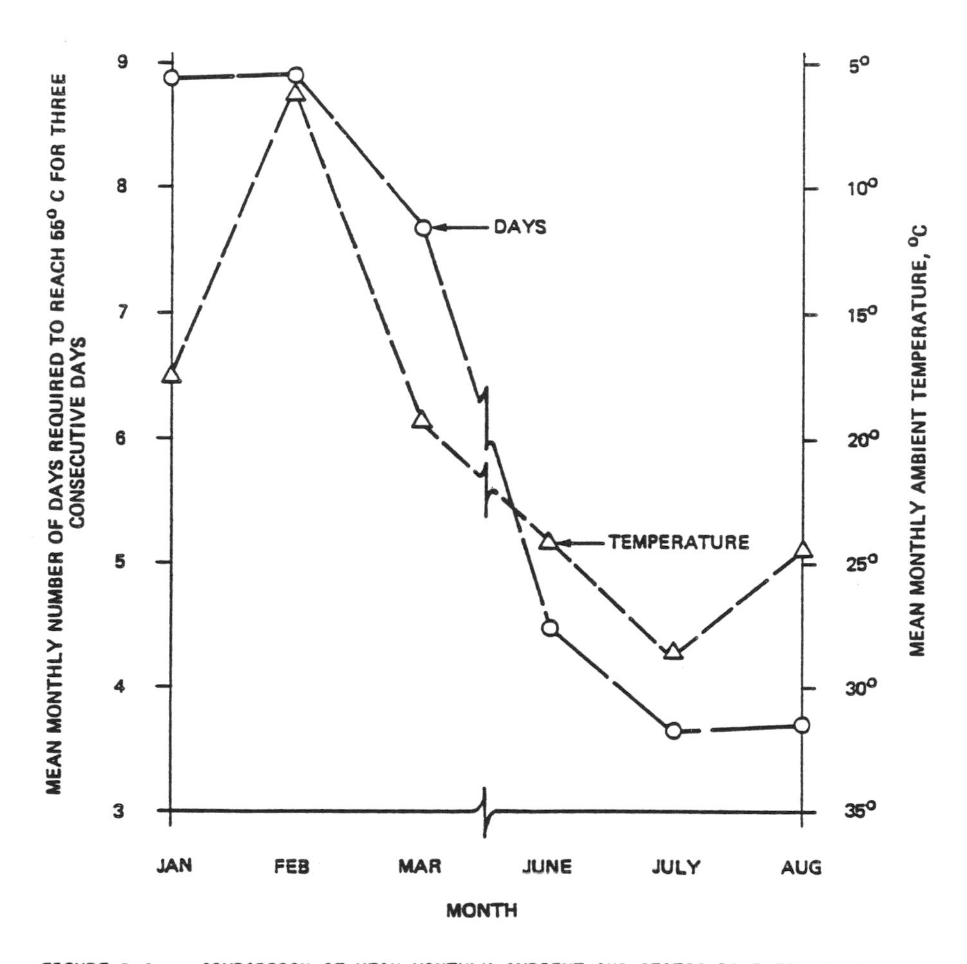

FIGURE B-1. COMPARISON OF MEAN MONTHLY AMBIENT AND STATIC PILE TEMPERATURES

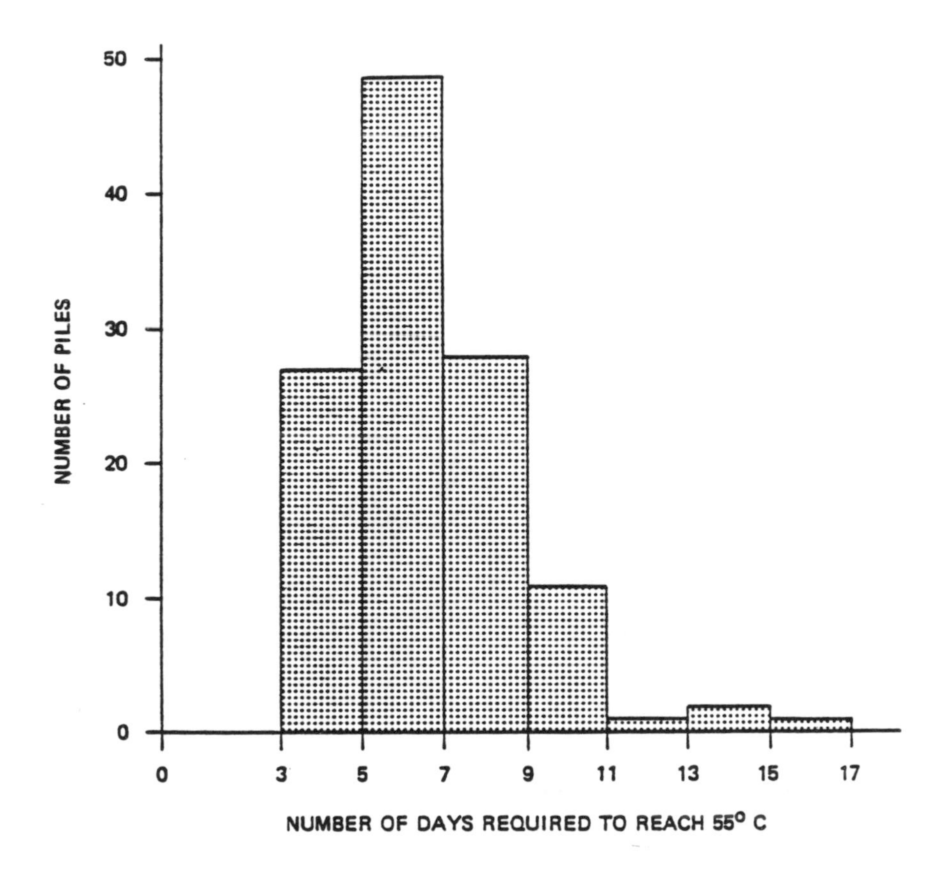

FIGURE B-2. FREQUENCY DISTRIBUTION FOR PILE TEMPERATURE DEVELOPMENT

The results of this analysis show that ambient temperature is an important determinant in static pile temperature development. The relationship between increasing ambient temperature and decreasing time to meet the static pile composting performance criterion of 55 degrees C for 3 days at the Hampton Roads facility is very strong as shown on Figure B-1.

Appendix C: Key Design and Operating Considerations for Static Pile and Windrow Technologies

TABLE C1

Item
Regulatory criteria[a]
Land disposal criteria
Time/temperature performance[b]
Composting period
Curing period
Process and/or ambient monitoring
Buffer zone requirements
Sludge haul requirements
Runoff control
Ambient air quality
Record keeping
Facility siting
Land availability and cost
Distance to treatment plant
Sludge haul route
Surrounding land use
Utility access
Site access
Treatment plant operations
Sludge type
Sludge production rate
Dewatering effectiveness
Dewatered sludge variability
Chemical additions
Dewatered sludge storage
In-plant coordination
Composting plant coordination
Sludge transport
Method
Distance and route
Delivery hours and frequency
Environmental impact
Equipment cleanup
Impact on sludge characteristics[c]
Sludge quantities
Planning period variability
Seasonal variability
Daily variability
On-site measurement
Sludge characteristics
Total and total volatile solids (moisture)
pH
Density
Heavy metal content
Nutrient content

[a]Federal, state and/or local.
[b]Process to further reduce pathogens (PFRP) or process to significantly reduce pathogens (PSRP).
[c]For example, density, free water, odor.
[d]Also applicable to aerated windrow.
[e]Applicable to conventional and aerated windrow.
[f]Static pile only.

TABLE C1. KEY DESIGN AND OPERATING CONSIDERATIONS FOR STATIC PILE AND WINDROW TECHNOLOGIES (continued)

Item
Sludge characteristics (continued)
Other constituents
Seasonal variability
Daily variability
On-site measurement
Site features
Site area requirements
Process area requirements
Paved operations
Covered operations
Drainage
Utilities
Support facilities
Materials handling logistics
Equipment, general
Type, by operation
Capacities, average and peak
Storage
Cleanup and maintenance
Process design, general
Time/temperature criteria
Volatile solids stabilization
Pathogen inactivation
Carbon-to-nitrogen ratio
Moisture control
Mix porosity
Oxygen content
pH
Heavy metal content
Operating areas
Process design, operations
Bulking agent/amendment use
Size and shape
Moisture content
Bulk density
Chemical characteristics
Availability and uniformity
Degradability
Recoverability
Mix ratio variability
Storage and materials handling requirements
Cost
Mixing operations
Mix moisture content
Mix uniformity and porosity
Sludge/amendment delivery

[a] Federal, state and/or local.
[b] Process to further reduce pathogens (PFRP) or process to significantly reduce pathogens (PSRP).
[c] For example, density, free water, odor.
[d] Also applicable to aerated windrow.
[e] Applicable to conventional and aerated windrow.
[f] Static pile only.

TABLE C1. KEY DESIGN AND OPERATING CONSIDERATIONS FOR STATIC PILE AND WINDROW TECHNOLOGIES (continued)

Item
Process design, operations (continued)
Equipment type, capacity, and operating flexibility
Time and motion requirements, including materials handling
Static pile construction
Dimensions
Base, cover material and placement
Time and motion requirements, including materials handling
Aeration pipe placement (with/without trenches)[d]
Aeration pipe type, material, and size[d]
Aeration pipe lateral configuration[d]
Aeration pipe perforation configuration[d]
Equipment requirements
Windrow formation
Dimensions
Shaping
Time and motion requirements, including materials handling
Access ways
Equipment requirements
Static pile aeration[d]
Mode, positive or negative
Rates(s) during active composting
Head loss (pipe and pile) and back pressure
Blower type, capacity, number per pile
Blower cycle and control method
Windrow aeration (and mixing)[e]
Equipment type and capacity
Turning frequency
Process monitoring
Temperature and oxygen
Equipment and method
Pile/windrow teardown
Equipment type and capacity
Environmental constraints (moisture, temperature, wind direction)
Time and motion requirements, including materials handling
Static pile drying
Minimum moisture content for screening
Alternative methods, aerated or unaerated
Equipment
Ambient influences
Time and motion requirements, including materials handling
Windrow drying[e]
Minimum moisture content
In-place drying time
Ambient influences

[a]Federal, state and/or local.
[b]Process to further reduce pathogens (PFRP) or process to significantly reduce pathogens (PSRP).
[c]For example, density, free water, odor.
[d]Also applicable to aerated windrow.
[e]Applicable to conventional and aerated windrow.
[f]Static pile only.

TABLE C1. KEY DESIGN AND OPERATING CONSIDERATIONS FOR STATIC PILE AND WINDROW TECHNOLOGIES (continued)

Item
Process design, operations (continued)
Screening[f]
Equipment type, capacity, flexibility
Moisture criteria for effectiveness
Screen size
Bulking agent recovery
Ease of cleaning
Mechanical reliability
Time and motion requirements, including materials handling
Curing
Screened or unscreened material
Method, aerated or unaerated
Time and motion requirements, includinig mateirals handling
Odor considerations
Potential sources
Sludge condition on arrival
Sludge accumulation on site
Condensate
High temperature
High moisture
Low oxygen content
Poor exhaust gas control
Ambient conditions conducive
Poor mixing
Poor housekeeping
Windrow surface area
Windrow turnings
Control methods
Compost scrubbers
Chemical scrubbers
Good compost practices
Changes in aeration
Aerosol considerations
Potential sources
Mixing
Active piles or windrows
Tear down
Screening
Dust generation
Control methods for operators
Enclosed equipment with filters
Filter masks
Control methods for public
Buffer zones
Water sprays
Wind direction and speed relative to siting

[a]Federal, state and/or local.
[b]Process to further reduce pathogens (PFRP) or process to significantly reduce pathogens (PSRP).
[c]For example, density, free water, odor.
[d]Also applicable to aerated windrow.
[e]Applicable to conventional and aerated windrow.
[f]Static pile only.

TABLE C1. KEY DESIGN AND OPERATING CONSIDERATIONS FOR STATIC PILE AND WINDROW TECHNOLOGIES (continued)

Item
Sidestream control
Potential sources
Leachate
Condensate
Runoff
Control methods
Operating procedures, leachate and condensate
Proper collection to prevent accumulations
Appropriate discharge
Finished compost quality
Chemical characteristics
Heavy metals
Plant nutrients
pH
Volatile solids
Biological characteristics
Pathogens
Carbon-to-nitrogen ratio
Physcial characteristics
Patricle size
Uniformity
Marketing and distribution
Cost
Regulatory constraints
Users
Distribution method
Private firm
Brokers
Direct public distribution, on site or delivery
Economics
Operating cost
Labor
Bulking agent/amendment
Aeration piping
Fuel
Power
Maintenance
Laboratory services
Sludge transport
Management, administration
Capital recovery
Land
Fixed facilities and equipment
Mobile equipment
Marketing revenue

[a]Federal, state and/or local.
[b]Process to further reduce pathogens (PFRP) or process to significantly reduce pathogens (PSRP).
[c]For example, density, free water, odor.
[d]Also applicable to aerated windrow.
[e]Applicable to conventional and aerated windrow.
[f]Static pile only.

Metric Conversion Factors

Multiply the U.S. customary unit			To obtain the SI unit	
Name	Symbol	by	Symbol	Name
Acceleration				
feet per second squared	ft/s^2	0.3048[a]	m/s^2	meters per second squared
inches per second squared	in/s^2	0.0254[a]	m/s^2	meters per second squared
Area				
acre	acre	0.4047	ha	hectare
acre	acre	4.0469×10^{-3}	km^2	square kilometer
square foot	ft^2	9.2903×10^{-2}	m^2	square meter
square inch	in^2	6.4516[a]	cm^2	square centimeter
square mile	mi^2	2.5900	km^2	square kilometer
square yard	yd^2	0.8361	m^2	square meter
Energy				
British thermal unit	Btu	1.0551	kJ	kilojoule
foot-pound (force)	$ft \cdot lb_f$	1.3558	J	joule
horsepower-hour	hp·h	2.6845	MJ	megajoule
kilowatt-hour	kW·h	3600[a]	kJ	kilojoule
kilowatt-hour	kW·h	3.600×10^{6}[a]	J	joule
watt-hour	W·h	3.600[a]	kJ	kilojoule
watt-second	W·s	1.000[a]	J	joule
Force				
pound force	lb_f	4.4482	N	newton
Flow rate				
cubic feet per second	ft^3/s	2.8317×10^{-2}	m^3/s	cubic meters per second
gallons per day	gal/d	4.3813×10^{-5}	L/s	liters per second
gallons per day	gal/d	3.7854×10^{-3}	m^3/d	cubic meters per day
gallons per minute	gal/min	6.3090×10^{-5}	m^3/s	cubic meters per second
gallons per minute	gal/min	6.3090×10^{-2}	L/s	liters per second
million gallons per day	Mgal/d	43.8126	L/s	liters per second
million gallons per day	Mgal/d	3.7854×10^{3}	m^3/d	cubic meters per day
million gallons per day	Mgal/d	4.3813×10^{-2}	m^3/s	cubic meters per second
Length				
foot	ft	0.3048[a]	m	meter
inch	in	2.54[a]	cm	centimeter
inch	in	0.0254[a]	m	meter
inch	in	25.4[a]	mm	millimeter
mile	mi	1.6093	km	kilometer
yard	yd	0.9144[a]	m	meter
Mass				
ounce	oz	28.3495	g	gram
pound	lb	4.5359×10^{2}	g	gram
pound	lb	0.4536	kg	kilogram
ton (short: 2000 lb)	ton	0.9072	Mg (metric ton)	megagram (10^3 kilogram)
ton (long: 2240 lb)	ton	1.0160	Mg (metric ton)	megagram (10^3 kilogram)

[a]Indicates exact conversion.

(continued)

METRIC CONVERSION FACTORS (continued)

Multiply the U.S. customary unit			To obtain the SI unit	
Name	**Symbol**	**by**	**Symbol**	**Name**
Power				
British thermal units per second	Btu/s	1.0551	kW	kilowatt
foot-pounds (force) per second	$ft \cdot lb_f/s$	1.3558	W	watt
horsepower	hp	0.7457	kW	kilowatt
Pressure (force/area)				
atmosphere (standard)	atm	1.0133×10^2	kPa (kN/m^2)	kilopascal (kilonewtons per square meter)
inches of mercury (60 degrees F)	in Hg (60 °F)	3.3768×10^3	Pa (N/m^2)	pascal (newtons per square meter)
inches of water (60 degrees F)	in H_2O (60 °F)	2.4884×10^2	Pa (N/m^2)	pascal (newtons per square meter)
pounds (force) per square foot	lb_f/ft^2	47.8803	Pa (N/m^2)	pascal (newtons per square meter)
pounds (force) per square inch	lb_f/in^2	6.8948×10^3	Pa (N/m^2)	pascal (newtons per square meter)
pounds (force) per square inch	lb_f/in^2	6.8948	kPa (kN/m^2)	kilopascal (kilonewtons per square meter)
Temperature				
degrees Fahrenheit	°F	0.555 (°F - 32)	°C	degrees Celsius (centrigrade)
degrees Fahrenheit	°F	0.555 (°F + 459.67)	°K	degrees kelvin
Velocity				
feet per second	ft/s	0.3048[a]	m/s	meters per second
miles per hour	mi/h	4.4704×10^{-1}[a]	m/s	kilometers per second
Volume				
acre-foot	acre-ft	1.2335×10^3	m^3	cubic meter
cubic foot	ft^3	28.3168	L	liter
cubic foot	ft^3	2.8317×10^{-2}	m^3	cubic meter
cubic inch	in^3	16.3871	cm^3	cubic centimeter
cubic yard	yd^3	0.7646	m^3	cubic meter
gallon	gal	3.7854×10^{-3}	m^3	cubic meter
gallon	gal	3.7854	L	liter
ounce (U.S. fluid)	oz (U.S. fluid)	2.9573×10^{-2}	L	liter

[a] Indicates exact conversion.

Other Noyes Publications

DRUM HANDLING MANUAL FOR HAZARDOUS WASTE SITES

by

K. Wagner, R. Wetzel, H. Bryson, C. Furman, A. Wickline, V. Hodge
JRB Associates

Pollution Technology Review No. 143

This book is a technical guidance manual on planning and implementing safe and cost-effective response actions applicable to hazardous waste sites containing drums.

The manual provides detailed technical guidance on methods, procedures, and equipment suitable for removing drummed wastes. Information is included on locating buried drums; excavation and onsite transfer; drum staging, opening, and sampling; waste consolidation; and temporary storage and shipping.

Each of these operations is discussed in terms of the equipment and procedures used in carrying out specific activities, health and safety procedures, measures for protecting the environment and public welfare, and factors affecting costs. Information is also included on the applications and limitations of several remedial measures for controlling or containing migration of wastes: surface capping, surface water controls, groundwater pumping, subsurface drains, slurry walls, and in-situ treatment techniques.

This manual will be useful to on-scene coordinators; federal, state, and local officials; and private firms that plan and implement response actions at sites containing drums.

CONTENTS

ISBN 0-8155-1121-3 (1987) **177 pages**